COMPUTER TECHNIQUES in
PRECLINICAL *and*
CLINICAL DRUG
DEVELOPMENT

T0174311

ROBERT C. JACKSON

CRC Press
Taylor & Francis Group
Boca Raton London New York

CRC Press is an imprint of the
Taylor & Francis Group, an **informa** business

Computer Techniques in

Preclinical *and* Clinical Drug Development

CRC Press
Taylor & Francis Group
6000 Broken Sound Parkway NW, Suite 300
Boca Raton, FL 33487-2742

First issued in paperback 2019

ISBN-13: 978-0-8493-7682-5 (hbk)
ISBN-13: 978-0-367-40128-3 (pbk)

Visit the Taylor & Francis Web site at
http://www.taylorandfrancis.com

and the CRC Press Web site at
http://www.crcpress.com

The Author

Robert Jackson graduated from the University of Cambridge (B.A. in Biochemistry, 1965) and the University of London (Ph.D. in Biochemistry, 1968). He did postdoctoral research in the United States on Fulbright-Hays scholarships at Yale University School of Medicine (Department of Pharmacology) and Scripps Clinic and Research Foundation (Department of Biochemistry); he then became a staff scientist at the Institute of Cancer Research, London and Surrey, England. He later held Assistant Professor and Associate Professor of Experimental Oncology positions at Indiana University School of Medicine.

His pharmaceutical industry positions held include Director of the Tumor Biology Section and Director of the Chemotherapy Department for Warner-Lambert/Parke-Davis Pharmaceutical Research Division; Group Director of Cancer Chemotherapy for DuPont; and since 1991, Vice President, Research and Development, for Agouron Pharmaceuticals, Inc., San Diego, California. He has participated in the discovery and development of over 20 compounds that have advanced to clinical trials, some of which are now marketed drugs.

Dr. Jackson served as member and chairman of the Experimental Therapeutics Study Section, Division of Research Grants, National Institutes of Health, and as a member of the Board of Scientific Counselors for the Division of Cancer Treatment, National Cancer Institute. He is author of one previous book and over a hundred research papers, reviews, and book chapters dealing with the pharmacology of anticancer drugs.

Preface

Of investigational drugs that enter clinical trials, it is variously estimated that 80 to 90% fail to reach the marketplace (DiMasi et al., 1994; Struck, 1994), dropping out of development because of insufficient clinical activity, unacceptable toxicity, rapid appearance of drug resistance, or related factors that should, in principle, be at least partially predictable from preclinical test systems. This work addresses the questions how can we improve the "hit rate" of developmental drugs and, in particular, can computational methods increase the predictive value of preclinical data?

In recent years the intensive use of computational methods for drug *discovery* has become widely accepted. These tools include molecular dynamics programs for study of atomic and molecular trajectories, which can be used to study energetics and molecular conformations. A variety of programs are available for the study of quantitative structure-activity relationships (QSAR) of potential drug molecules. "Docking" programs are widely used to predict which ligands have the potential to bind to particular receptor sites. At a much more experimental stage are programs that attempt to predict binding affinities of potential inhibitors to their target macromolecules. Molecular graphics programs that make it possible to visualize, in atomic detail, structures of receptor and enzyme proteins and their interactions with substrates and inhibitors are widely available and have made a major contribution to our ability to visualize the subtleties of drug-target interactions. None of these computational aids to drug design are discussed in the present volume. Readers needing an entry to the computational chemistry literature are referred to the annual review series edited by Lipkowitz and Boyd (1990 and later).

Computational tools are becoming increasingly available for study of most of the critical decision points in drug *development*. These tools are of two kinds: data analysis techniques, which provide statistical and model-fitting end points for interpretation of preclinical and clinical experiments, and modeling techniques, which attempt to describe complex biological systems in mathematical terms and which can be used for experimental design and for study of ways in which the subcomponents of a complex system may interact. This book discusses tools of the second kind. It is not intended as a technical treatise on modeling, but as a guide to existing methods and how to use them in making drug development decisions.

The design and analysis of clinical trials is a specialized branch of biostatistics that is not discussed in depth here; the use of computer programs that simulate clinical trials is considered in Chapter 8, but for a detailed discussion of biostatistics, readers are referred to the standard texts (e.g., Cox and Snell, 1981; Bailar and Mosteller, 1992).

Typically, drug development decisions are made at a time when relatively little is known about the new compound; only after the compound has become an established drug are the details of its biochemistry, pharmacology, and toxicology filled in. The essence of drug development, therefore, is to make intelligent predictions, often from minimal data. The established techniques of enzyme kinetics, pharmacokinetics, pharmacodynamics, and cytokinetics provide powerful tools for making generalizations from limited experimental data. Newer techniques, such as knowledge-based expert systems, neural nets, and simulations of complex biochemical systems, are beginning to make a contribution to the prediction of drug metabolism and toxicity.

The aim of the present work is both descriptive and prescriptive. In the first capacity, it reviews the currently available computational tools for making drug development decisions, discusses their strengths and limitations, and explores some of the new directions in computational pharmacology. Over 30 commercially available programs are discussed. It is concluded that drug development is becoming increasingly computer-intensive, but that presently the field is fragmented and consists of a variety of unrelated design tools. In a more prescriptive mode, this book considers how a modest investment in additional preclinical experimental work, coupled with much more intensive computational analysis, should enable more reliable predictions of clinical activity to be made from the available preclinical data. Ultimately this will require integrated computational systems that can intercommunicate, access large knowledge bases, and extract the greatest possible information content from expensive and limited experimental work.

In fact, the accumulation and manipulation of large chemical and biological data bases is another existing specialized area that the present work, with its emphasis on pharmacological modeling, is unable to cover in depth, though this subject will be touched upon in Chapter 4. In many companies the data-base management is the function of a specialized group, often termed Research Information Systems (RIS). Pharmaceutical companies are accumulating a great wealth of information in these data bases, though they generally do a better job of acquiring this information than of exploring its depth and power.

It is the author's belief that to the three areas where computer technology already makes a major contribution to drug design and development (computational chemistry, biostatistics, and data base management) a fourth will be added — computational pharmacology. The other clear trend is toward greater integration of these technologies: computational chemistry techniques for calculating physical properties of drugs (log P, solubility, etc.) are clearly relevant to an understanding of drug absorption and distribution; biostatistical techniques calculate the pharmacokinetic parameters used for modeling studies, and large data bases contain within their depths the knowledge base that can form the basis for the expert systems that will play an increasingly important role in drug development decisions.

The author is grateful to many colleagues at Agouron Pharmaceuticals, Inc. for discussion of the ideas presented in this book, but particularly to Ted Boritzki and Bhasker Shetty. The author would also like to acknowledge with thanks the help and guidance of Marsha Baker at CRC Press, Inc.

Table of Contents

Abbreviations

ADME study	absorption, distribution, metabolism and elimination study
AICAR	5-aminoimidazole-4-carboxamide ribonucleotide
AICARFT	5-aminoimidazole-4-carboxamide ribonucleotide formyltransferase
araC	arabinosylcytosine (cytarabine)
AUC	area under the curve (concentration–time integral)
ATDA	2-amino-1,3,4-thiadiazole
AZT	azidothymidine (zidovudine)
BBB	blood–brain barrier
BCNU	bis-(chloroethyl)nitrosourea (carmustine)
BRM	biological response modifier
Cl_p	plasma clearance
CMC	critical micelle concentration
CNS	central nervous system
CPE	cytopathic effect
CSF	cerebrospinal fluid
CTL	cytotoxic lymphocytes
dFdCTP	2′,2′-difluorodeoxycytidine-5′-triphosphate
DHFR	dihydrofolate reductase
EC_{50}	50% effective concentration
F%	percent oral availability
FABP	folic acid binding protein
FDA	United States Food and Drug Administration
FdUMP	5-fluorodeoxyuridylate
FPGS	folylpolyglutamate synthetase
GAR	glycinamide ribonucleotide
GARFT	glycinamide ribonucleotide formyltransferase
GI	gastrointestinal
HGPRT	hypoxanthine-guanine phosphoribosyltransferase
HIV	human immunodeficiency virus (AIDS virus)
HPLC	high performance liquid chromatography
IC_{50} (or IC_{90})	50% (90%) inhibitory concentration
IP	intraperitoneal
IV	intravenous
K_d	dissociation constant for receptor-ligand complex
KEM	kinetic effect model
K_i	dissociation constant for enzyme-inhibitor complex
K_m	Michaelis constant of an enzyme
LD_{10} (or LD_{50})	10% (50%) lethal dose
log P	log of n-octanol–water partition coefficient
LSD	lysergic acid diethylamide
MIC	minimum inhibitory concentration
MLP	maximum life span potential

MoAb	monoclonal antibody
6-MP	6-mercaptopurine
MTD	maximum tolerated dose
MTX	methotrexate
NDA	New Drug Application (to FDA)
PALA	N-phosphonacetyl-L-aspartate
PB-PK model	physiologically-based pharmacokinetic model
PD	pharmacodynamic
PK	pharmacokinetic
PK/PD model	combined pharmacokinetic and pharmacodynamic model
QSAR	quantitative structure-activity relationship
QSPR	quantitative structure-pharmacokinetic relationship
QSTR	quantitative structure-toxicity relationship
RFC	reduced folate carrier
SCID	severe combined immunodeficiency
$t_{1/2}$	plasma half-life
TMTX	trimetrexate
TS	thymidylate synthase
Vd	volume of distribution
V_{max}	maximal velocity of an enzyme

CHAPTER 1

Introduction:
The Preclinical and Clinical Drug Development Process

I. THE HEURISTIC APPROACH TO DRUG DEVELOPMENT

A former colleague of the author used to maintain that to make a development decision for a new antibacterial, it was necessary to know two things about the compound: its MIC and the achievable blood level. *MIC* means minimum inhibitory concentration, i.e., the lowest concentration of the inhibitor that gives complete suppression of visible growth in a broth dilution assay (Pratt and Fekety, 1986). This is just one of the hundreds of "rules" that constitute drug development folk-wisdom. Some other examples are (1) no compound with a molecular weight over 1200 will make it as a drug (this one predates the use of recombinant human proteins, though not the use of insulin), (2) to cross the blood–brain barrier a compound has to have log P more positive than -1.5 (log P is the log of the n-octanol–water partition coefficient), and (3) all nitroaromatics are toxic.

Concerning these "rules", two conclusions come to mind: first, they encapsulate a lot of hard-earned experience and common sense; second, it is possible to think of exceptions to all of them. Let us consider the first example a little more carefully. A more conservative version of the statement would be to recast it in a negative form: if an antibacterial cannot achieve a blood level higher than its MIC, it is unlikely to be any good. Most pharmacologists and microbiologists would agree with that, but some drugs are rapidly eliminated into the urine, so that the urine concentration greatly exceeds the blood level, and if you are trying to develop an antibiotic for urinary tract infection, the blood level may not be as important as the urine level. Again, for a topical antibacterial the blood level may never get very high, but what matters is the level in the infected skin at the site of application. So what we really mean is, if an antibacterial cannot achieve a blood level (or a concentration in the infected organ or tissue) higher than its MIC, it is unlikely to be any good. This is theoretically better, but of less practical help, because drug levels in many of the potential sites of interest may be harder to measure than the blood level in experimental animals and impossible to measure in humans. Also, what do we mean by blood (or tissue) level: peak level, minimum level, or mean level? Ideally, we should like to know all these things. Since, however, in real life, we cannot know the concentration of our drug everywhere in the body at every time-point after treatment, we have to resort to taking samples and extrapolating or interpolating from our sample data. To continue the example of a developmental antibiotic, we might treat human volunteers, take samples of blood and urine at hourly intervals after treatment, calculate plasma half-life and other pharmacokinetic parameters, and supplement this information with a study of tissue distribution in rats.

It would be interesting to collect the rules of thumb (*heuristics* is the fashionable word) that drug developers use and subject them all to this kind of critical deconstruction. However, the point that emerges from such analysis is that once a rule gets more complicated than two or three pieces of information, it starts to become a quantitative exercise, and we need a quantitative description (or model) to which to fit the data and statistical procedures to analyze the results.

The author suspects that a good deal of drug development gets done "by the seat of the pants." Recently, a potential development partner (for one of the experimental antiviral agents of the author's company) described its requirements for a drug as follows: it should be able to maintain a 99% inhibitory concentration for at least 8 h after a human dose of not greater than 500 mg, it should have a plasma half-life of not less than 2 h, etc. When asked where these requirements came from, the answer was, "a consensus opinion of our development team." In fact, these were not unreasonable requirements, but the trouble with this "seat of the pants" approach was that it did not help with the question, How much worse would the drug be if its half-life were only 90 min? In short, it was not quantitative.

Modeling approaches in this field as in any other, allow one to do the *Gedankenexperiment*: What if the peak plasma level were *x* instead of *y*? What if the plasma kinetics were biexponential instead of monoexponential? If the pathogen becomes fivefold resistant, how is drug efficacy affected? This book discusses the modeling techniques that are available for studing questions of this sort. It considers the related issue of how much (or how little) experimental data is necessary to make reasonable development decisions. Reverting to our hypothetical antibiotic, if we know the MIC and the peak blood level, and if we could know one more piece of information, what should that third piece of data be?

II. *IN VITRO* AND *IN VIVO* MODELS OF DRUG ACTIVITY

If we are ethically to take a potential new drug into human trials for the first time, we must, as a minimum, have some reason for believing that it might work and that it will be safe. In fact, we generally know quite a lot more about the compound.

The various stages of preclinical and clinical drug development are discussed in detail below. Let us first consider the requirement that we must have some reason for believing the compound might work.

A. Receptor-Binding Kinetics

In the early days of drug development, it was quite common for drugs to be developed without a knowledge of the molecule's mechanism of action. For example, if a compound displayed a low MIC value against bacteria and gave high blood levels in rodents without toxicity, it stood a fair chance of becoming a useful antibacterial drug and might well be on the market for many years before the molecular site of action was characterized. While there are occasional examples of this happening nowadays, the current emphasis on "rational drug design" has made this much less common. Thus, usually the first piece of information we have about a candidate drug molecule (other than chemical information) is its activity in (depending on the class of drug) a receptor-binding assay or an enzyme-inhibition assay. The result is typically expressed as an IC_{50}, i.e., the concentration that gives 50% inhibition of an enzyme or displaces 50% of the natural ligand from a receptor (often, alternatively, termed the EC_{50}, the median effective concentration). It needs to be remembered that the IC_{50} value is not necessarily constant. A more rigorous end point is the K_i (for an enzyme) or K_d (for a receptor), which are true dissociation constants. (Methods of calculating these values are discussed in Chapter 5.) A modest amount of extra work at this stage will often provide valuable additional information, such as whether the inhibitor is competitive or noncompetitive with the natural ligand and what the apparent reaction order is (from the Hill coefficient).

B. *In Vitro* Efficacy Assays

If the experimental agent is being tested as an antibacterial compound, it is usually possible to determine what concentration and exposure time are necessary for an antibacterial effect against bacteria growing in a nutrient broth in the test tube and whether the agent is bacteriostatic or bactericidal. The usual end point for antibacterial agents is the MIC (defined above). At this point, if the compound is active, it is also customary to collect some preliminary information on the agent's **spectrum**, i.e., the range of pathogens against which it is active. *In vitro* tests will often be used to determine the compound's **cross-resistance profile**, i.e., whether the new compound is fully active against bacterial isolates with acquired resistance to existing drugs. It is also possible to explore the new compound's **selectivity**; i.e., by examining the effect of the compound on human cells growing in culture, we can determine whether it is a nonselective poison, indiscriminately toxic to all living cells, or whether it has selective antibacterial activity. For the latter to be of any practical use, we would expect a candidate antibacterial drug to be at least 100-fold (and normally 1000-fold) more toxic to bacteria than to human cells.

Antiviral testing is similar in principle, though more complicated technically, because a host cell must be provided for the virus to grow in. Since the virus will eventually lyse the host cells, a test with an active antiviral agent will increase the cell count of an infected culture. Studies against noninfected host cells can be done to control for possible cytotoxicity of the inhibitor to host cells. The 50% effective concentration (EC_{50}) is the concentration that gives 50% protection of the host cell population.

The first stage of anticancer testing is normally done *in vitro*, often against mouse tumor cells, but more commonly in recent years against human tumor cells. The usual end point is the IC_{50}, the concentration that decreases the treated cell count by 50% in comparison with an untreated cell culture, after some arbitrary period of time (often 48 or 72 h). An IC_{50} is not a constant, and its value will be highly dependent upon the test conditions, especially the length of the drug exposure. Similarly, it is possible to define an IC_{90} value that gives 90% inhibition of tumor cell proliferation, and the ratio of the IC_{90} to the IC_{50} will give a measure of the steepness of the dose–response curve (see Chapter 5). As with antibacterial agents, it is possible to distinguish between cytostatic and cytotoxic agents. In the case of anticancer drugs, *spectrum* refers to the range of tumor types in which the compound is active, so frequently new compounds will be evaluated against established cell lines from lung, breast, and colon carcinomas, as well as against lymphoma or leukemia cells and other tumor types. While this is quite easy to do, it is not yet clear how accurately the *in vitro* spectrum of an anticancer agent correlates with its eventual clinical spectrum. This subject will be discussed further below. The cross-resistance profile of potential antitumor agents can also be explored *in vitro*, and test laboratories commonly maintain cell lines that are resistant by various known mechanisms to the clinically useful anticancer drugs, as well as to agents known to be chemically or mechanistically related to the test compound.

Some investigators believe that the biology of established cell lines has diverged so far from the *in vivo* situation that their drug sensitivity no longer reflects that of the original tumors. If this is the case, it is desirable to study drug activity against primary tumor material. The human tumor cloning assay is aften used for this purpose. It certainly appears that cells from primary tumor biopsies are generally less drug-responsive than established cell lines. Such tests are often run against a panel of different tumor types, using a 1-h drug exposure (or longer exposure times where the drug effect is known to be time dependent), often using one tenth of the peak plasma level (measured in mice) as a starting point, and scoring as a positive result tests in which the colony count is reduced by a predetermined amount, usually either 50 or 70%. The statistics collected for the human tumor cloning assay suggest that the predictive value for negative results is about 86%, i.e., if the particular tumor type does not respond *in vitro*, the compound is unlikely to be clinically active against that tumor type. The predictive power for positive results was less (64%), but still significant (Von Hoff et al., 1983). The human tumor cloning assay thus represents a way of predicting an anticancer drug's clinical spectrum from preclinical data.

Selectivity of anticancer agents is harder to study *in vitro*, because many normal cell types do not grow in culture, or if they do, they are arguably no longer normal. This situation is expected to change as we learn more about the growth factor requirements of the various normal cell populations. In the meantime, bone marrow cells are widely used as a normal tissue comparison (Parchment et al., 1994), because the bone marrow is a dose-limiting site of toxicity for many anticancer drugs, and the growth factor requirements of bone marrow are well understood and the factors in many cases are commercially available. When IC_{50} values are known for both tumor cells and bone marrow cells, it is possible to calculate an *in vitro* therapeutic index, defined as the ratio of bone marrow IC_{50} to tumor cell IC_{50}. For a drug whose dose-limiting toxicity *in vivo* is myelosuppression, the *in vitro* therapeutic index should give a rough estimate of the safety margin.

Where it is possible to do cell culture assays, as in the case of antibacterial, antiviral, and anticancer drugs, these tests can provide a wealth of information that could not be obtained from studies with an isolated enzyme or receptor. In particular, if the target is an intracellular one, it will be necessary for the test compound to penetrate the cell membrane. In many cases where a compound shows potent binding to its target molecule, but then shows poor activity or no activity against intact cells, lack of adequate transport across the cell membrane is responsible. In other cases, rapid intracellular metabolism of the test compound may be the culprit.

The usefulness of cell-based assays is not confined to development of drugs for cancer and infectious disease. In other diseases, even though the pathological process may be a complex process involving many cell types, if the particular drug target under study is known to be located in a particular cell type, it may be possible to devise a cell-based assay. For example, the target molecules for many immunosuppressive and anti-inflammatory drugs are located within T-lymphocytes, so that cell-based assays using T-cells may provide valuable test data.

C. *In Vivo* Models of Drug Activity

The art and science of experimental pathology is concerned with creating artificial disease states in test animals. These can then be used as models in which experimental drugs, eventually intended for use in humans, may be tested. It is readily apparent that this general approach raises two questions. The first concerns the species difference: will the biochemical, anatomical, or physiological differences between the rat (or other test species) and the human affect drug response? The second question is, to what extent does the artificially induced disease process resemble the natural disease? These two questions encapsulate the whole uncertainty of drug development, since, obviously, if any *in vivo* model were totally predictive then clinical trials would be merely a formality and pharmaceutical research would be less of a lottery and more of an accounting exercise.

Given the limited predictive value of *in vivo* models, and the fact that their application is the central problem of drug development, it would be fatuous to attempt to discuss the whole question of the utility of animal models, in all therapeutic areas, in the current work. The question we have to consider here is, how can computer techniques be used to reduce some of the uncertainties of drug development? In this connection it may be useful to repeat a few of the more widely held generalizations about animal models. First, their predictive value for clinical activity varies greatly between therapeutic areas; at one end of the scale we have antibacterial drugs, where, given that the same organism that we are trying to eradicate in humans (or a closely related organism) will grow in a test species, if we can kill the bacterium in a mouse, other things being equal, we can kill the same bacterium in a human. At the other end of the scale of predictive value we have mental illnesses such as psychosis, depression, or senile dementia. A rat that has been subjected to certain torments may, within limits, behave in a way that resembles these human conditions, and treatment with experimental drugs may alleviate the condition, but we are clearly dealing with uncertainties far greater than those involved in treating a bacterial infection. In between these two extremes are hypertension, cancer, diabetes, and a host of other conditions where experimental diseases in animals can be used to test drugs, so long as the limits of applicability of the model system are kept constantly in mind. The second major generalization is that optimal dosage, time and route of elimination, and the rate (even the route) of drug metabolism may be species dependent. To the extent that the species differences are understood, we can attempt to extrapolate from the animal experience to what we may expect to find in humans. The way in which this is done forms a major part of the content of the present work. The third generalization is that the predictive value of an *in vivo* model is enormously increased if we have a biochemical end point, especially if the same biochemical end point can be used to follow the progress of clinical trials. This is possible, but not always so; we can follow the progress of treatment of prostate cancer in rats or humans by measuring a marker protein (alkaline phosphatase or prostate-specific antigen) in the blood. On the other hand, measuring amyloid plaque deposition in brain as an end point in a rat model of Alzheimer's disease may be more objective than behavioral tests, but it would be neither ethical nor practical to do repeated brain biopsies in a human patient.

A final comment about *in vivo* model systems is that this area of research is being revolutionized by genetic engineering technology. The most obvious example is the use of transgenic mice, i.e., mice into whose genome a foreign gene has been stably transfected. If a mouse carries a human gene that codes for a drug target protein, assuming that protein is functional, then from the viewpoint of drugs that act on that particular enzyme or receptor, the transgenic animal may be considered as a model of the human situation (it may or may not be necessary to disable the mouse's own homolog of the gene under study). Other genetically engineered mice are useful for particular purposes: mice with severe combined immunodeficiency (SCID mice) have no immune system, because they have no functional B or T lymphocytes. If these mice are "reconstituted" with human B and T cells, the resulting SCID/Hu mice now have a human immune system. They may thus be used as models for study of diseases of the immune system, or of drugs that act as immunosuppressants or immunostimulators, with the knowledge that many of the uncertainties caused by species differences have been eliminated.

A very familiar model system in cancer research is the xenograft (which means a foreign graft), particularly of a human tumor growing in an immune-deficient mouse. While thymectomized, irradiated mice may be used for this purpose, it is common practice to use nude mice, congenitally athymic mice that also happen to be hairless (Giovanella and Foght, 1985). Despite their lack of an immune system, these animals are quite hardy if kept away from sources of infection, and many human tumors will grow in them. These animals may thus be used for testing the effect, *in vivo*, of experimental anticancer drugs on human tumors. Of course, the therapeutic effect of an anticancer drug represents a balance between the drug effect on the tumor and its effect on the normal tissues. Since the normal tissues are those of a mouse, and mouse tissues tend to differ systematically from their human counterparts in drug sensitivity (e.g., the bone marrow tends to be more resistant to many anticancer drugs than that of humans, but the gastrointestinal epithelium less so), the nude mouse carrying a transplanted human tumor cannot simply be regarded as a miniature human; allowance must still be made for species differences.

D. Why Are Preclinical Models Not More Predictive?

We have already referred to the fact that preclinical model systems have two basic limitations: there may be species differences that confuse interpretation of results (e.g., a particular drug may be metabolized much more rapidly in humans than in mice) and the pathology of the model system may not be an exact approximation of the human disease. Sometimes there may be a major species difference in the target macromolecule; e.g., a current approach to developing drugs against arthritis is to design inhibitors of fibroblast collagenase. However, rats, the most common test species for inflammation testing, do not have a fibroblast collagenase. Often the factors are more subtle; e.g., mice tend to greatly overpredict for efficacy of drugs that block cell proliferation by inhibiting the *de novo* pyrimidine biosynthetic pathway, such as N-phosphonacetyl-L-aspartate (PALA) and pyrazofurin, because mice have much lower circulating uridine concentrations than humans. Since uridine can be salvaged by uridine kinase and partially circumvent the inhibition of the *de novo* pathway, human plasma much more effectively neutralizes such drug effects than mouse plasma. A converse situation is seen for thymidylate synthase inhibitors, such as Tomudex: mouse plasma contains high concentrations of thymidine, which tends to block the effect of thymidylate synthase inhibitors, while human plasma has much lower levels. Thus, mice tend to underpredict for toxicity and efficacy of this class of compounds.

Other examples of species differences in toxicity are known where, independently of efficacy, a class of compounds may be much more or less toxic to rodents than to humans; this may shift the therapeutic index either way. Sometimes these species differences can be attributed to different sizes of stem cell compartments; e.g., mice have proportionately more bone marrow stem cells than humans, but fewer gastrointestinal crypt stem cells. This tends to make cytotoxic drugs less myelosuppressive in mice than in humans, but more gut toxic. In anticancer testing, simple size differences between species may have a surprisingly large effect. The experimental tumors used for drug testing in mice will often have a mass of 50 to 300 mg at the time of the test. A human tumor, at the time of treatment, is more likely to be in the range of a few grams to a few hundred grams, greatly complicating the drug delivery problem. Another factor that predisposes mouse tumors to be more drug sensitive than human tumors is the generally differing cytokinetics of tumor cells in the two species. In general, murine tumors have shorter cell cycle times, lower cell loss factors, and smaller quiescent fractions, all factors that correlate with drug sensitivity.

Obviously, the uncertainties of extrapolating from *in vitro* models to the clinical situation are even greater. Apart from the obvious limitations of *in vitro* systems, the lack of drug delivery concerns and selectivity issues, there may be quite unexpected complications. In general, if a particular concentration of penicillin kills a particular strain of Streptococcus *in vitro*, the same concentration will kill the same organism *in vivo*. However, in the case of human tumor cells, there is evidence that the three-dimensional organization of a tissue may alter the drug sensitivity of the component cells manyfold. The same cells, grown *in vitro* on reconstituted extracellular matrix, may give a response much closer to the *in vivo* situation than when they are grown as monolayers on plastic or glass.

This discussion of why preclinical models are not more predictive has dwelt on some of the factors that may cause preclinical test systems to either overpredict or underpredict the clinical activity of anticancer drugs. An analogous list could be produced for any therapeutic category. Even though the premise of the present work is that some of these limitations may be overcome by more sophisticated computer models, clearly, intelligent use of such models requires an understanding of the biology of the human disease and its animal model and a knowledge of the comparative pharmacology of the drug class under study in humans and test animals. Using any model system, whether an animal model or a computer model, without such knowledge can result in some extremely expensive (and sometimes tragic) mistakes.

E. What Can We Do to Make Preclinical Models More Predictive?

Perhaps the single greatest confounding factor in predicting likely clinical results from preclinical data is the species difference. Wherever possible, using human cells or tissues will be preferable to using rodent material. Since it is often desirable to use an *in vitro/in vivo* model (e.g., to use cells of a particular tumor for *in vitro* studies and to grow the same tumor *in vivo* for experimental chemotherapy), *in vitro* studies in both rodent and human cells will be necessary. To continue the example of anticancer drug development, it is now fairly common practice to use *in vitro/in vivo* systems consisting of the same established human tumor line grown both in tissue culture and as a xenograft in nude mice. This is probably more predictive of the likely clinical response than using a murine tumor, but the host body is still that of a mouse. Using transgenic mice or SCID/Hu mice so that the desired target macromolecule is human rather than murine will increase the predictive power of the preclinical data, but such systems are not yet available for testing most drug classes, and at present transgenic animals are used more to study the underlying biology and pathology of disease than for testing new drugs. A recent highly publicized example of the use of transgenic mice as models of human disease is the development of a mouse carrying the gene for human β-amyloid precursor protein and the suggestion that this could be a test system for drugs for Alzheimer's disease.

We can improve the predictive power of preclinical test systems by acquiring a little more biochemical information about the disease, the target macromolecule, and the drug. The example was mentioned above of drugs that inhibit *de novo* pyrimidine biosynthesis. Several such compounds were taken to clinical trials

(6-azauridine, PALA, pyrazofurin, and brequinar) before the species difference in plasma uridine levels between mice and humans was discovered. In this case, knowing the biochemical problem would not necessarily have suggested a way of solving the problem, but it might have saved money on clinical trials that were doomed or modified the clinical trial design in such a way as to minimize the antagonism of these compounds by uridine in human plasma.

We can also improve the predictive value of preclinical data by maximizing the usefulness of that data through more intensive computer-assisted analysis. If we know that drug X inhibits enzyme Y, and that humans have tenfold less of enzyme Y than mice do, then we can create a pharmacodynamic model that will examine what happens to drug efficacy and toxicity when we change the activity of enzyme Y. Such a model will doubtless involve some uncertainties, but its predictions should give drug developers a much better idea what to expect than trusting to luck or intuition. If drug Z is known to be myelosuppressive, if this myelosuppression is time dependent, and we suspect that (as if often the case) the drug will be eliminated more slowly in humans than in mice, how much should we adjust the phase I starting dose to allow for this potentially greater toxicity? A combined pharmacokinetic/pharmacodynamic model can help us in this situation. Discussion of such computer models to get maximum mileage out of preclinical data, how to use them, and what kinds of information they can provide, is the primary theme of the succeeding chapters of this book.

III. PRECLINICAL PHARMACOKINETICS

One of the primary pieces of information that we need about a new drug candidate before taking it to human trials is knowledge of its rates of uptake, distribution, and elimination. If our *in vitro* studies have shown that a drug candidate starts to influence its receptor at 100 nM, and is approaching its maximal effect at 5 μM, then one of the most basic things we need to know about that compound is whether we can reach that blood level without seeing unacceptable toxicity. For many drug types, it is important to know whether the compound can be given orally. The first information of this kind usually comes from the preclinical pharmacokinetic (PK) study. The preclinical pharmacokinetics of most drugs is generally studied in rats.

A distinction must be made between classical pharmacokinetics and the more recent (and labor-intensive) approach of physiologically based pharmacokinetics (PB-PK). PB-PK models are discussed in detail in Chapter 2. The present discussion focuses on classical pharmacokinetics. The classical approach is to treat the body as a black box; drug enters and leaves the body in quantities and at rates that can be precisely quantitated, but what goes on in the box can only be inferred. Conventionally, the body is treated as a small number of compartments, between which drug is transferred through first order kinetics. In PB-PK, by contrast, the various tissues and organs of the body (or a subset of interest) are explicitly modeled. The PB-PK approach generates powerful models, but requires much more experimental data and computer power to be used effectively.

It is not the function of the present book to describe pharmacokinetic theory. Of the many good available books on this subject, that of Gibaldi and Perrier (1982) is regarded as a standard work, and the present author also makes frequent use of Welling (1986) and Wagner (1993). The preclinical PK workup of a new drug candidate first requires development of an assay for measurement of the drug in rat plasma or serum. While many analytical techniques may be used for this purpose, high-pressure liquid chromatography (HPLC) is the most generally used method. Often the plasma assay developed for the preclinical work can later be used to measure drug in human plasma during the clinical PK study. Based upon the time-course of blood levels, the data are generally fitted to a two- or three-compartment open model and plasma half-lives for the various phases are calculated, along with volume of distribution (V_d), area under the plasma concentration–time curve, or AUC (often called $C \times T$ by chemotherapists), and plasma clearance, Cl_p (usually defined as dose/AUC). If the drug is being developed as an oral dosage form, it is possible that an intravenous formulation may not exist, in which case all the PK information may be obtained from orally administered drug. In this case, the peak plasma level and the time from drug administration to the plasma peak are additional values that would usually be measured. If both oral and intravenous formulations are available, it is usually of interest to determine the oral bioavailability (F%) in rats, i.e., the fraction of an orally administered dose that reaches the bloodstream. This can be calculated from the ratio of AUC values after oral and intravenous administration of similar doses.

A. Tissue Distribution

Synthesis of a batch of radiolabeled drug is usually a fairly expensive proposition, so it is not usually feasible to make radiolabeled batches of three or four rival compounds to see how they compare. The radiolabeled synthesis is one of the activities that generally takes place once the decision has been made to move a compound into preclinical development. Usually this material will be carbon-14 labeled, though tritiated material is sometimes used. Ideally, several batches labeled at a variety of sites would be the best way of exploring a compound's metabolism, but this is not generally feasible — a single radiolabeled derivative is usually all one can afford. For this reason, careful thought has to go into where to put the label. For example, if a drug is an acetyl ester, it would probably be easy synthetically to put on a carbon-14 acetyl group. However, this would almost certainly be rapidly removed by plasma and tissue esterases, so that no information would be obtained about the subsequent fate of the drug. The first experiment done with the labeled drug when it becomes available is the absorption, distribution, metabolism, and excretion (ADME) study in rats. In fact, the ADME study is primarily a time-course of tissue distribution; it usually provides the first glimpse of where the candidate drug goes in the body. Does it accumulate in adipose tissue? Does it penetrate the blood–brain barrier? How high are the liver levels? How much of the drug is eliminated in urine? In the feces? The plasma and urine samples also provide an opportunity to search for drug metabolites; the fact that metabolites are formed and the rate at which they appear are easily established, though isolating and identifying the metabolites may be more difficult. It is not usually necessary to identify rat metabolites of the drug before beginning clinical trials, though it is generally expected that the major human metabolites will have been identified before approval for marketing is granted.

B. Toxicology

As we noted above, before we can ethically test a drug in humans, we need to have some reason for believing it will be effective and for believing it will be safe. The first of these requirements is the purpose of the preclinical pharmacology, discussed above. Meeting the second requirement is the function of the toxicology studies. These studies attempt to answer the following questions: is the test compound toxic? If so, at what dose? Are there any clinical signs of toxicity? Do the blood chemistry tests or the red blood cell count or white blood cell count show changes after administration of the test substance? After necropsy of the treated animals, are drug-induced changes seen in body organs, either visibly or as seen by histopathological examination?

Typically, toxicity is studied in one rodent species (usually rats) and one nonrodent species (generally beagle dogs). The exact design and duration of the toxicology studies will depend on the therapeutic category of the drug and also on the design and duration of the planned clinical trial. The usual expectation is that before exposing humans to the drug, the toxicology animals will have been exposed to higher doses (the greater the safety factor the better) and for a treatment duration at least as long as the planned phase I clinical trial duration. The design of toxicology studies for anticancer drugs, for example, is discussed by Grieshaber (1991). For many drug classes, though usually not for anticancer drugs, it is customary to test for genotoxicity before beginning clinical trials. The usual battery of tests includes an Ames test, which examines the mutagenic potential of the test compound in bacteria, and tests for chromosomal aberrations *in vitro* and in the bone marrow of treated animals.

For some drug classes it is also customary, before commencement of clinical trials, to conduct two additional batteries of tests, the cardiovascular and central nervous system safety pharmacology. These studies determine whether the test compound causes changes in cardiovascular or CNS function: Are potentially dangerous changes seen in blood pressure or heart rate? Is the compound a CNS depressant, etc.?

After clinical trials of a drug have commenced, further animal toxicology studies are usually conducted. The purposes of these additional studies is to examine the safety of longer-term exposure, to obtain further information about toxic effects that may have been noted in the clinical studies, and to obtain information on possible carcinogenicity and reproductive toxicity of the test compound. These studies may take years to complete, at a cost that can run into millions of dollars, as well as the lives of hundreds of laboratory animals. It sometimes seems that toxicology protocols are less a scientific investigation with clearly defined objectives than a mindless ritual enacted to propitiate the regulatory authorities, who in turn find it easier to demand more data than to make decisions based on an objective balance of risk and benefit. Clearly, the issue of drug safety is one of the foremost concerns of the drug development process, and though risk can never be eliminated from the process, it must be minimized. Equally clearly, the question of what constitutes an acceptable risk is not a purely scientific one, but involves more fuzzy issues of society's values.

C. Other Preclinical Development Activities

After the decision has been made to move an interesting compound into development (i.e., when the preclinical pharmacology studies have given encouraging results), it typically takes 1 to 2 years before the company is ready to start clinical trials. The activities that have to take place during that time include, first of all, chemical scale-up, often to the 1-kg scale at this point, to provide material for toxicology and for the phase I clinical trial, and radiolabeled synthesis, to provide material for the ADME study. While this is going on, analytical chemists will be developing an analytical method that can be used for monitoring drug stability and possibly can also be used for the PK studies. The pharmaceutical development group will be examining the compound for solubility, stability, and compatibility with likely excipients and container materials, with the objective of producing a formulation, which may be intravenous, or oral, or some more specialized dosage form. Once sufficient drug is available, the toxicology (which is generally the rate-limiting step) can begin.

Obviously, not all compounds that move into preclinical development will advance to clinical trials. Sometimes the toxicology or safety pharmacology studies will uncover an unacceptable toxicity; sometimes it may be impossible to formulate a compound so that it has acceptable solubility or stability. While companies tend not to discuss their failures, so that reliable data are hard to collect, the attrition rate of candidate compounds during preclinical development has been estimated to be about 50%.

D. Computer Models as a Substitute for Animal Research

It is often claimed that computer models can substitute for animal testing. This may sometimes be the case, but the issue is complex. The position of the present author is that drugs could be developed with much less animal testing than is currently the case, by substituting more *in vitro* toxicology and by making more extensive use of modeling techniques, but that an irreducible minimum amount of animal research will remain. Let us assume that an *in vitro* toxicology assay has shown that a test compound is safe below 10 μM, but causes a dose-dependent and time-dependent cardiomyopathy above that level. How do we determine the significance of this observation? Clearly, we must relate it to the achievable peak *in vivo* dose (or, in some cases, to the AUC) and to do this *in vivo* PK data are necessary. If we must do animal testing, and the author believes we must, we have a responsibility to see that these experiments give the maximum possible amount of information.

The use of computer models to integrate *in vitro* efficacy and toxicity data with *in vivo* PK information is one way to accomplish this objective.

IV. THE PURPOSE OF PHASE I CLINICAL TRIALS

Toward the end of the preclinical development period, the clinical research team will write a clinical protocol for a phase I trial. The phase I protocol will propose a starting dose, the schedule of administration, and the clinical and laboratory tests that will be conducted on the subjects. It will discuss eligibility criteria for the subjects and the criteria that will be adopted for deciding when it is safe to proceed to a higher dose. It will also state what end points will be used to decide when to terminate the trial and how the phase I results will be used to decide upon a protocol for the phase II clinical trial. This phase I clinical protocol, together with the preclinical pharmacology report, the pharmacokinetics and toxicology reports, and the chemistry and manufacturing reports, comprises the initial submission to the regulatory authority. In the United States this document is termed an Investigational New Drug Application, or IND. When the regulatory authority (in the United States this is the Food and Drug Administration [FDA]) gives its permission, the phase I clinical trial can begin. The purposes of a phase I trial are to get some indication about what is likely to be an effective dose in humans, to find out what, if any, are the toxicities caused in humans by the compound, and to learn something about its pharmacokinetics in humans. Since the purpose of a phase I trial is not to establish efficacy, it is often conducted in healthy volunteers. In the case of some life-threatening diseases, such as cancers, where relatively toxic compounds may be administered, it is considered unethical to administer these test compounds to healthy volunteers, so in such cases the phase I study will be done in an appropriate patient population.

The initial dose used for the phase I clinical trial is based on the preclinical toxicology. Often the starting dose (scaled on a milligram per kilogram body weight or milligram per square meter body surface area basis) will be one tenth of the maximum tolerated dose in mice, or one tenth of the highest "no effect" level in dogs, if that is lower. If this low dose is tolerated without toxicity by the first three phase I volunteers, the dose will be increased, progressively, until either toxicity is seen or a target blood level (e.g., for an antibacterial drug, some multiple of the MIC) is reached. While the dose is being escalated, blood samples from the volunteers are being analyzed. The phase I study thus gives information on human pharmacokinetics, such as peak blood level, plasma half-life, AUC, etc. After this single-dose information has been obtained, the study may be extended to a multi-dose phase, which will provide information on, for example, whether the blood level will accumulate if the compound is given every day for a week.

The phase I volunteers are medically examined frequently for possible toxicity. They are also interviewed for information on problems such as headaches, nausea, or mental symptoms. These subjective problems may be unexpected, because the animal toxicology does not give any information on how the animals feel. Typically toxic symptoms are graded on a scale of zero to four. The severity of toxicity that would prevent further dose escalation will vary, depending upon the disease indication. In the case of an anticancer drug study, a toxic effect of grade three or four seen in more than one subject at a particular dose level would be regarded as indicating that the maximum tolerated dose had been reached. In addition to clinical signs of toxicity, toxicity is also scored based upon blood counts, blood chemistry, and other laboratory tests.

The typical attrition rate for drugs in phase I is unknown. Of the 20 experimental agents whose development the author has participated in, 3 failed to make it to phase II, in all cases because of unacceptable toxicity. This proportion of phase I drop-outs is probably typical, but the author has not seen any confirming statistics.

V. THE PHASE II CLINICAL TRIAL

The purpose of a phase II trial is to find out whether the candidate drug works in humans. The trial design will vary according to the therapeutic category, but the initial objective is to study a relatively small number of patients (often 12 to 25) with a fixed dose and schedule of administration and to look for signs of activity. For many classes of drugs, it is customary to do a placebo-controlled trial. For example, one may be performed for a new agent being tested as a topical treatment for psoriasis, since this is a condition where the lesions may regress spontaneously or the supposedly inactive ingredients in the formulation may have a healing effect, independent of the drug. Thus, a placebo, identical in appearance and composition, with the single exception of not containing the test substance, can control for spontaneous remission or for vehicle effects. These studies are often double-blind — neither the doctor nor the patient knows whether the treatment being given to a particular individual is the placebo or the test substance. In the case of life-threatening diseases, such as cancer or AIDS, phase II trials are not usually placebo controlled. At the end of the initial phase II study, the results will usually be subjected to statistical analysis. However, it is inherent in the study design of early phase II studies, and the small number of patients who participate in them, that they do not have much statistical power. By the end of this phase of clinical testing, one should have enough information to tell whether the compound is active in humans. Generally, one will not have any reliable information on comparative activity. How does the compound compare with other drugs in its therapeutic class? To get reliable information on response rates, and on the duration of the responses, requires larger patient numbers. For some drugs, an expanded phase II clinical trial may provide sufficient data to file a New Drug Application (NDA) with the FDA for permission to market the drug. Since the FDA expects to see proof of efficacy in two independent studies, and to achieve sufficient statistical power will probably require a minimum of about 100 patients in each study, it will require clinical experience in over 200 patients (including those in phase I and the early phase II study) to get a drug to market. Usually the numbers are larger than this, often reaching into the thousands, especially if more than one dose or schedule of administration (e.g., once daily vs. twice a day) is being explored. Whether a drug is likely to be approved based upon phase II data depends upon the urgency of the medical need, which, in turn, is often related to what alternative drugs are available for the particular therapeutic condition and how effective they are. Of the last ten anticancer agents to receive marketing approval in the United States, five were approved based upon expanded phase II data.

Phase II clinical trials are the single most critical hurdle in drug development, because this is where we find out whether the drug works in humans or not. It is at this stage where most of the attrition occurs; while estimates of phase II failure rates vary, almost everyone agrees that fewer than 50% of compounds complete phase II successfully, and a success rate of one third is probably closer to the usual experience. From the economic perspective, one function of phase II trials is to abandon inactive compounds with a minimum expenditure of time and money.

VI. THE PHASE III CLINICAL TRIAL

Across the whole range of therapeutic indications, most drugs require phase III data to receive marketing approval. The essential difference between a phase II study and a phase III study is that a phase III trial is a randomized, comparative study. Another difference is that in a phase III trial, the test compound may be employed in combination with other drugs. For example, in a phase III clinical trial of an anticancer agent, we may evaluate activity against metastatic breast cancer by comparing response rates of patients treated with a standard drug combination (e.g., cytoxan, Adriamycin, and 5-fluorouracil) with a combination in which the test drug had been substituted for Adriamycin. When a new patient is enrolled, she would be assigned to a particular arm of the study (control arm or test arm) by the study coordinator. While a two-arm study is the simplest phase III study design, multiarm studies, in which more than one dose of the test substance or more than one kind of combination is studied, are quite common. Even more complex designs may be used, such as the crossover study, in which certain events may result in a patient being switched from one arm to another. For example, if a patient did not respond within a reasonable period of time to the test arm (or ceased to respond after an initial response) she might be "crossed over" to the control arm. Part of the reason for the complexity of phase III trial designs is that the investigators wish to get a clear indication of the level of efficacy of the test drug without compromising the right of the patients to receive appropriate and effective treatment.

Phase III trials are time-consuming and expensive. It may take months or years to accrue the large number of patients needed, and more years of patient follow-up before the study is completed. Because of the expense and complexity of phase III trials, biostatisticians are heavily involved in their design, with the responsibility of suggesting a trial design that will answer the question to a predetermined level of statistical power in a minimum amount of time and with the smallest necessary number of patients (making reasonable allowance for likely dropouts). The present work makes no attempt to review the literature on biostatistics and clinical trials design, though there are, of course, program packages available to assist with these important functions. Statistics packages range from large, well-maintained systems such as SAS and BMDP to PC-based programs such as Statistica and MultiStat.

While the design and analysis of clinical trials is emphatically not a job for amateur statisticians, these packages provide a means for drug developers to "get a quick idea" of how many patients it would probably take to get an answer to a given degree of statistical significance for a particular trial design. The Clinical Trials Design Program, developed at the Johns Hopkins Oncology Center, is a commercially available, PC-based program that performs sample-size and power calculations for a variety of biomedical study designs using established literature methods (see the Appendix for information on software availability).

VII. PRECLINICAL AND CLINICAL END POINTS

The most fundamental end point in a clinical trial is a clinical measure of the pathological process that is being treated: does an antihypertensive agent reduce blood pressure? Does an anticonvulsant drug prevent, or decrease, the frequency of epileptic seizures? Does an anticancer agent shrink tumors? If the disease being treated is a life-threatening condition — heart failure, cancer, AIDS, then the ultimate measure of the drug's success is survival, and clinical trials of drugs for these indications will usually ask the question of whether, in comparison with the existing standard treatment, the new compound increases survival. This can take many years, and, for reasons both of ethics and economics, if the new compound is less effective than the standard treatment, it is desirable to terminate the trial as soon as possible; hence the interest in surrogate markers, i.e., clinical or laboratory tests that are expected to predict for survival. In cancer, partial or complete responses (tumor shrinkage or apparently total disappearance) often correlate with a survival advantage, though for some tumor types it is not clear whether a moderate degree of tumor shrinkage necessarily always results in longer survival. In AIDS, a decrease in circulating virus RNA, or an increase in the CD4 lymphocyte count, indicate that the body has been partially cleared of the virus, and that the immune system is improving, and it is expected that these surrogate markers should correlate with longer survival. Of course, if this correlation were an established one, they would not be surrogate markers any more; they could be considered to be a legitimate end point.

An increasing number of biochemical and immunological diagnostic tests are becoming available for monitoring the progress of treatment. Two frequently used tests in oncology are for carcinoembryonic antigen (CEA) as a marker for colon carcinoma and prostate-specific antigen (PSA) as a marker for prostate cancer. These tests (and a large number of other well-studied diagnostics, in a wide range of diseases) have the advantage of being rapid, objective, and quantitative. They can give an early indication of whether the disease is responding to treatment, and how rapidly, and how completely; they can give early warning of relapse. These *in vitro* diagnostics can sometimes be used both in test animals and in human patients, and in such cases a comparison of the human and animal results can give valuable feedback on the predictive value of the animal model.

There are a number of noninvasive physical techniques that can be used as end points, such as computerized tomography scans, positron emission tomography, and nuclear magnetic resonance (NMR) imaging, that can provide a very graphic picture of diseased internal organs and their response to treatment. In a few cases these techniques can be adapted to the direct measurement of drug levels in the tissues and organs. For example, fluorine NMR can be used to measure accumulation of the anticancer drug 5-fluorouracil and its metabolites. Since it is generally very difficult to get measurements of drug levels in human tissues, these studies provide a valuable way of validating the predictions of human drug distribution that are made from preclinical data.

VIII. A COMMENT ON DEVELOPMENT COSTS AND TIMELINES

There have been a number of recent estimates (by drug companies and by the Pharmaceutical Manufacturers Association) that a new drug costs in the range of $250 million to $400 million to develop to the marketplace and that the process typically takes 9 to 12 years. A contrary view has been expressed by some academic economists, who suggest that between $12 million and $100 million is more realistic. This 30-fold discrepancy may be due to several factors: first, the low figure of $12 million assumes that the drug is a breakthrough product in a therapeutic category where it would receive rapid regulatory approval based on clinical trials in perhaps 300 patients, where no toxicity concerns necessitated additional toxicology studies, and where there were no major scale-up, formulation, or manufacturing problems to be overcome; a best case, in other words. For a more typical drug, where clinical trials might require 2000 patients, and where there are usually multiple chemical and toxicological hurdles requiring attention, the figure of $90 million would not be unreasonable. Both these figures ignore the cost of failed development candidates; it is not unheard of for a drug to be taken all the way to NDA filing, yet never receive FDA approval. Obviously, these "dry holes" have to be paid for. In other words, if we take a typical drug company's annual R&D budget, and divide it by the average number of approved NDAs per year, we come up with a figure of around $250 million. If we also factor in the return that this money would have produced if we had invested it in government savings bonds, or some other safe and steady interest-paying investment, we are now close to the $400 million figure, which, it may be argued, is the real economic cost of drug development. Now the author is proposing that we increase that mind-boggling sum still further by doing additional research. This can only be justified if we can improve the success rate; more realistically, the author argues that the additional studies proposed here, and the additional computational analysis, will help us to identify our failures early, before we have spent too much money on them.

IX. WHY DO DEVELOPMENTAL DRUGS FAIL?

Drugs fall out of the development process for two main reasons: insufficient clinical activity and unexpected toxicity (i.e., toxicity that was not anticipated by the preclinical toxicology studies). A third, and less common, situation is when a drug is both safe and effective, but problems of chemical instability or difficulties of large-scale synthesis make it economically infeasible to bring it to market. This last reason for failure is beyond the scope of the present analysis: this work focuses on the first two factors and asks how we might do a better job of predicting drug safety and efficacy.

X. HOW CAN MATHEMATICAL MODELING IMPROVE THE SUCCESS RATE?

The remaining chapters discuss in detail five computational techniques that address this question and review the software tools that are currently available for this purpose. The five techniques are PB-PK modeling, PK/PD modeling, cytokinetic modeling, kinetic simulation of multienzyme pathways, and the use of expert systems for prediction of drug metabolism and toxicity. As a preview, listed below are seven examples of drug development problems where more intensive use of computational methods could provide insights that would help in decision-making.

1. An antibacterial drug was shown in an ADME study to concentrate in kidney to levels higher than the plasma concentration. To be effective against organisms causing bacterial nephritis it must be maintained at a concentration of greater than 1 μM for 96 h. What is the likely human dose?

    Comments: Classical pharmacokinetics should provide information on the dosage necessary to maintain a bacericidal plasma concentration for the required treatment duration. A PB-PK model (Chapter 2) would provide more useful predictions of actual tissue levels.

2. An antiemetic agent is believed to act on the chemoreceptor trigger zone of the brain. It has log P of –0.5, shows efficacy as an antiemetic at doses above 40 mg/kg in ferrets and dogs, and its peak brain level in a rat ADME study was 4% of the peak plasma level. Doses above 100 mg/kg cause severe hypertension in rats.

    Comments: This agent acts in the brain, but is has rather marginal ability to penetrate the blood–brain barrier. Up to a point this can be overcome by using higher doses, but this strategy is limited by what is probably a vasoconstrictive effect at high doses. Its usefulness will probably depend upon the ratio of brain level to blood level that can be expected in humans. Interspecies scaling predictions (Chapter 3) based upon a PB-PK model (Chapter 2) could provide some insights as to whether the therapeutic margin is large enough to be usable.

3. An amphetamine analogue is rapidly metabolized in rats by hydroxlation and glucuronidation, and its half-life is too short for it to be pharmacologically active. Are there other test animals that would be more appropriate, and what are the chances that the compound could have an acceptable half-life in humans?

    Comments: A pharmacologist experienced in drug metabolism studies would probably not need a computer to predict that guinea pigs would be a more appropriate model than rats for this class of drug; however, an expert system for prediction of drug metabolism (Chapter 4) could provide some quantitative predictions.

4. An antileukemic agent acts by inhibiting the enzyme cytidine 5′-triphosphate synthetase (CTP synthetase) and has strong activity against the L1210 leukemia in mice. A typical human lymphoblastic leukemia has activity of CTP synthetase

threefold lower than the L1210, and the K_i for inhibition of human enzyme is fivefold higher than for the mouse enzyme. What is the net effect of these differences likely to be?

Comments: As stated above, the problem is oversimplified: the total activity (V_{max}) of target enzyme and its K_i for inhibitor are only two of the factors that will influence response to the inhibitor in intact cells. Other factors are concentration of competing substrate (UTP in this case), activity of other enzymes that utilize the competing substrate, and activity of converging pathways that produce the product of the inhibited reaction (e.g., cytidine salvage). To model all these factors in mouse and human is possible with an appropriate pharmacodynamic model of the pyrimidine biochemical pathways (Chapter 5).

5. An experimental antitumor drug has given strong responses against a number of human colon carcinomas grown as xenografts in nude mice. The drug is known to act in the *S* phase of the cell cycle. While human tumor xenografts are an excellent model of the human disease in many respects, they generally have a high growth fraction (>50%), while a colon carcinoma growing *in situ* may have a growth fraction of <10%. How is this likely to affect tumor responsiveness?

Comments: One of the greatest limitations of human tumor xenografts is the fact that cytokinetically they tend to select for rapid proliferation, while a human tumor *in situ* has a complex biology that is influenced by host factors that are probably not present (or are at least different) in a mouse. One way to study this situation is to obtain cytokinetic parameters by studying primary human tumors (where possible) and then factor the observed parameters into a cytokinetic model (Chapter 6).

6. An experimental AIDS drug reaches blood levels in healthy human volunteers that correspond to an average inhibition of viral replication of 99.3%. The mutation rate for resistance to the drug is 10^{-4}, and the mutants are about 100-fold resistant. When the drug is tested in AIDS patients, how long is the response likely to last?

Comments: This is an interesting problem because it combines pharmacokinetics, cytokinetics, and the prediction of drug resistance (Chapter 7). If the response duration is considered too short to be of any practical use, it raises the question: how much does it help to increase the dose or the frequency of dosing?

7. If there are ten useful partially non-cross-resistant drugs available for the treatment of AIDS (as there soon will be), there are 5860 ways of combining four or fewer drugs. If each drug can be administered in four dose levels or frequencies of administration, there are over 1.3 million possible protocols. If we can afford to look at 20 of these in clinical trials, which should they be, and how do we choose them?

Comments: This is the well-known phenomenon of combinatorial explosion. Generally, combination protocol design already factors in a lot of preclinical and clinical information on cross-resistance patterns, metabolic interactions, and overlapping toxicity. However, this is a complex problem, and rather than attempting to optimize clinical protocols simply by experience and intuition, we need objective ways of building in as much knowledge about the disease and the drug as possible. The use of modeling tools for combination protocol design is discussed in Chapter 8.

CHAPTER 2

Physiological Pharmacokinetic Modeling

I. THE LIMITATIONS OF CLASSICAL PHARMACOKINETICS

Nobody would nowadays contemplate taking a new drug into clinical trials without some preclinical PK information — as a minimum, data on plasma half-life, clearance, and volume of distribution in rats. The value of such information is undisputed: we need to know what plasma concentration to expect (and this can often be related to *in vitro* efficacy data), how long this concentration will be sustained after a single dose, and what dose was required to achieve it.

One of the most common reasons for drugs dropping out of development is inability to deliver the required amount of drug to the target site for the required amount of time. This may be due to insufficient oral bioavailability, or (for a dermatological product) poor skin penetration, or to differences in PK or metabolism between test animals and human subjects. For those drugs that act primarily in the bloodstream, classical PK can tell us most of what we need to know. The classical PK approach has the advantages of simplicity, ease of validation, and applicability to clinical studies. It can provide information that may be used to estimate a starting dose for the initial human trials (Collins et al., 1990). Carefully interpreted, classical PK data can often provide additional hints about how a drug is being handled in the body; e.g., systemic clearance approximating the hepatic blood flow suggests that biliary excretion may be the primary route of drug elimination. A recent, comprehensive review of classical PK modeling techniques has been written by Bourne (1995).

There are a number of well-maintained and well-documented computer programs commercially available that can be used for classical PK analysis. Typically, these programs require the user to set up a spreadsheet of plasma drug levels at different times; the program will then calculate the usual PK parameters (with estimates of the error and other statistical information) and also provide graphical output. These programs can then be used for modeling, so that once the classical PK parameters are known, we can predict the blood levels that result from given doses and schedules of administration. With an anti-infective drug, for example, we often want to know how large a dose is required to maintain trough levels between doses that are some safe multiple of the MIC. We may wish to calculate the consequences of switching from four-times-daily to three-times-daily dosing, and so forth. Programs that can readily provide this kind of information include MKMODEL, PCNONLIN, and P-PHARM. In all cases, a choice of several predefined models is available (usually one- and two-compartment exponential models, with either i.v. bolus, i.v. infusion, or oral drug input), and user-defined models can also be run. These classical PK modeling programs are established drug development tools and need not be discussed in further detail here. Information on the availability of these programs is provided in the Appendix. Despite its undoubted usefulness, however, classical PK provides no information about drug levels in particular tissues, and for many drugs the primary target will lie in some tissue or organ other than blood. If we are to analyze distribution of such drugs to their target site in humans and animals, it is clear that classical PK cannot help us.

II. PHYSIOLOGICALLY BASED PHARMACOKINETIC MODELS

As discussed in Chapter 1, an ADME study with radiolabeled drug, generally in rats, is part of the normal preclinical workup for a developmental drug. For an anticancer drug, the studies can be done in a rat bearing a transplanted solid tumor. These studies provide a first look (and sometimes the only look) at where the drug is going in the body. They also give information on the time-course of accumulation and elimination for each of the organs and tissues examined. After the ADME study we are in a position to answer such questions as: what is the peak ratio of drug concentration in liver relative to plasma? Does the drug cross the blood–brain barrier? Does it show slow accumulation in adipose tissue? An ADME study with enough time points contains all the information we need to provide a detailed description of drug distribution (though often an ADME study is conducted at a single dose, and, ideally, to validate a quantitative model we would like information at a number of drug doses). However, there are many quantitative questions that can only be answered if we have fitted a quantitative model to the ADME data. Consider the following questions: (1) We have an antibiotic that we wish to use to control bacterial endocarditis. How much drug will ensure that the level in the heart will not fall below 5 μM? (2) A drug is neurotoxic if the brain level rises above 1 μM, but safe below that level. How soon after an initial dose is it safe to administer a second dose? (3) Tumor cells *in vivo* mutate to a methotrexate-resistant form in which the V_{max} for membrane transport is decreased threefold. How will this affect tumor levels of drug after a 10-mg/kg i.v. bolus dose?

To model questions of this kind, we need a physiological pharmacokinetic model (often described as a physiologically based pharmacokinetic model, or PB-PK model). Credit for the development of this technique is generally attributed to Dedrick and Bischoff (see Bischoff and Brown, 1966; Bischoff and Dedrick, 1968; Dedrick et al., 1970). Bischoff (1986) traced some of the underlying ideas back to work done by T. Teorell in the 1930s, but observes that Teorell was unable to develop his ideas because of lack of computer power. In a PB-PK model, drug is considered to partition into compartments that represent actual organs and tissues. The amount of drug entering and leaving each compartment will depend upon the blood flow rate to each organ, the diffusion or transport of drug between blood and tissue, and the partition factor (relative affinity) of drug between blood and tissue. A schematic diagram of a PB-PK model is shown in Figure 2.1. This is an 11-compartment model. Arterial blood is considered to flow to eight organs; venous blood from these organs passes through the lungs and is then recirculated. Rate Q_{17} of the model represents the hepatic portal circulation. Rate Q_{18} represents biliary secretion of drug from the liver into the gut lumen, from which it may be eliminated into feces, or reabsorbed, thus completing an enterohepatic circulation. Drug may enter the system by three routes: intravenously (either as an instantaneous bolus or as an infusion over time), orally, or as an intrathecal infusion directly into the brain. A blood–brain barrier is considered to exist that impedes both the uptake of drug from the blood into the brain and loss of drug from the brain. Intrathecal infusion avoids the blood–brain barrier. Drug may leave the system by three routes: by excretion from the kidney into the urine, by elimination from the gut lumen into the feces, or by metabolic conversion in the liver.

Physiological pharmacokinetic models have now been described for a growing number of drugs (which remains, however, a minority of all useful drugs). Some helpful background reading on the subject can be found in the article by Bischoff (1986), and in chapters of the books by Welling (1986) and Wagner (1993). The advantages and disadvantages of the PB-PK approach have been succinctly summarized by Welling (1986), and the following list is based upon his discussion. Advantages of the technique are:

1. Free and bound drug concentration–time profiles can be predicted for any organ of interest.
2. The PK parameters are based upon measurable physiological values (organ weight, tissue blood flow, etc.) rather than upon mathematical abstractions such as volume of distribution.
3. Parameter values can be altered to reflect physiological changes; e.g., if tumor blood flow is decreased (because of intratumoral pressure or the effects of an antiangiogenic drug), we can factor this into our model and predict the consequences for tumor accumulation of, for example, a cytotoxic drug.
4. Specific organ elimination mechanisms can be modeled; e.g., if a particular drug is eliminated by both renal and biliary excretion and the ratio of the two varies under different physiological conditions, a PB-PK model (unlike a classical PK model) could predict the total elimination correctly.
5. Parameter values can be scaled for different species, so that if we have a good PB-PK model of a particular drug in rats, we can substitute the human physiological parameters into the model, and this will give us (as a first approximation) a model of the disposition of that drug in humans. The topic of interspecies scaling is discussed in detail in Chapter 3.

As disadvantages of PB-PK modeling, Welling (1986) lists the following:

1. Model development requires more experimental data than is required for a classical model.
2. PB-PK models are mathematically more complex than classical PK models.

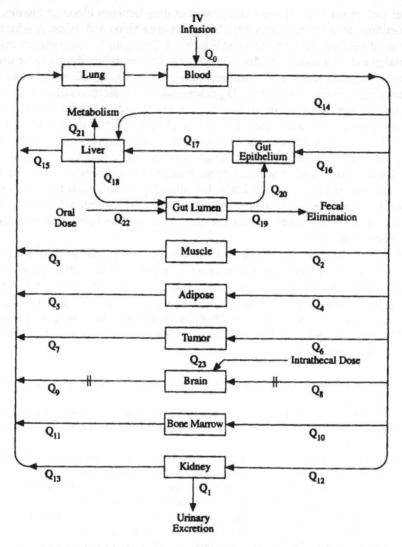

Figure 2.1 Schematic diagram of an 11-compartment PB-PK model. The flow rates between compartments are indicated by Q_0 to Q_{23}. Three routes of drug administration are possible (oral, i.v., and intrathecal), and three routes of drug elimination are modeled.

3. PB-PK models are harder to validate experimentally in animals (and virtually impossible in humans) because large numbers of tissue samples are required at multiple time points after dosing.
4. It is difficult to obtain tissue samples that are free of blood.
5. *In vitro* testing is frequently required to establish or validate PB-PK model parameters (e.g., drug partition constants for the various compartments may have to be measured *in vitro*).
6. Despite their complexity, PB-PK models still make many oversimplified assumptions about diffusion of drug into tissues, partitioning within organs, and the relative degree of intravascular and extravascular drug binding.

This last comment is somewhat contradictory of the earlier ones; having complained that PB-PK models are complex, we now complain that they are still too simplistic. This disadvantage can be partially offset by collecting more data and making the model more complex. In fact, PK models form a continuum from the highly oversimplified (but easily validated) single-compartment classical model to quite realistic mathematical descriptions that may require years of work to construct and validate. Where do we draw the line? Historically, drug developers have drawn the line at classical multiexponential PK models and left anything more complex to academic investigators. The thesis of the present work is that, because of advances in analytical technique, and in computer hardware and software, it is now becoming feasible to develop a PB-PK model as part of the routine preclinical workup for developmental drugs.

Mathematically, a PB-PK model is a system of differential equations. These equations are mass-balance equations, one for each compartment, describing rates of appearance and loss of drug into and out of each compartment as a function of time. The equations for these models are given by Welling (1986) and Wagner (1993) and are briefly summarized in the next section. PB-PK model-simulating programs are available, such as the author's PHYSIOPLEX, for solving the equations. Many authors use general-purpose simulation programs, such as SimuSolv, SIMULINK, or STELLA II to construct PB-PK models (see Appendix).

A. Mass Balances

For a typical model such as that of Figure 2.1, the amount of drug entering (or leaving) tissue compartment i per unit time is given by the difference in drug concentration between arterial (C_a) and venous blood (C_v) multiplied by the blood flow to the compartment, Q_i, so the change in drug concentration in the compartment, C_i, is this value divided by the compartment volume, V_i:

$$dC_i/dt = Q_i/V_i \cdot (C_a - C_v) \tag{2.1}$$

If the partition factor between tissue i and blood is R_i then

$$C_v = C_i/R_i \tag{2.2}$$

If the fraction of the drug that is free (unbound) in blood is f_B then the fraction of free drug in tissue i is given by

$$f_i = f_B/R_i \tag{2.3}$$

Combining Equations 2.1 and 2.2 gives us

$$dC_i/dt = Q_i/V_i \cdot (C_a - C_i/R_i) \tag{2.4}$$

If this quantity is positive, there is net uptake of drug by the tissue from blood; if dC_i/dt is negative then there is net loss of drug from the tissue to the blood.

For a compartment from which drug elimination occurs, such as kidney or liver, an additional term is necessary to describe the elimination; e.g., if this is assumed to be a first order process and is a function of free drug concentration in the kidney (or liver) then the rate of elimination from tissue i is

$$r_i = k_i \cdot f_i \cdot C_i \qquad (2.5)$$

For hepatic metabolism, r_i will usually be saturable and will be calculated by the Michaelis–Menten equation or by some more complex function. The total rate of change for the elimination tissue is

$$dC_i/dt = [Q_i/V_i \cdot (C_a - C_i/R_i)] - r_i \qquad (2.6)$$

Conversely, the change of drug concentration in the blood is given by the sum of the individual tissue contributions:

$$dC_a/dt = \sum_{i=1}^{i=n} Q_i \cdot C_i/R_i - Q_B \cdot C_a \qquad (2.7)$$

where n is the number of compartments, other than blood, and Q_B is total blood flow through the n modeled compartments.

B. An Example of a Flow Limited PB-PK Model

An early example of a PB-PK model was that of Harris and Gross (1975) for Adriamycin kinetics in rabbits. This is a flow-limited model, i.e., it assumes that drug equilibrates rapidly between blood and tissue, so that the amount of drug in a particular tissue at a particular time depends on the plasma concentration, on the partitioning between plasma and tissue, and on the blood flow to and from that tissue (for some drugs this assumption does not hold, and these transport-limited models are discussed below). The Harris and Gross (1975) model included plasma, bone marrow, liver, lung, heart, "lean tissue" (muscle), kidney, gut, spleen, and adipose tissue compartments. This choice of compartments for modeling reflects the kind of information we need from the model. Kidney, being the major site of elimination for many drugs, must obviously be included. For Adriamycin, hepatic metabolism is a major source of drug elimination, so this is also an essential compartment to model. The "lean tissue" compartment represents about 45% of total body mass in a range of species, including mice and humans, so it is too important to omit by virtue of its sheer bulk. Adipose tissue often accumulates significant amounts of drug and, in the case of fat-soluble drugs, may act as a slowly accumulating, slowly releasing reservoir compartment. Bone marrow and gut represent the major sites of acute toxicity for many anticancer agents, including Adriamycin, so it is important that they be included. Heart was of particular interest because Adriamycin can cause a severe chronic cardiomyopathy whose incidence is related to the lifetime cumulative dose. The model predicted that after an i.v. bolus injection of 3 mg/kg of

Adriamycin, all eight solid tissue compartments would rapidly (within a few minutes) accumulate higher drug concentrations than the plasma; at 24 h, levels ranged from 27 times the plasma level in liver to 450 times the plasma level in kidney. This high drug concentration in all of the tissues examined is related to the high volume of distribution found for Adriamycin in classical PK modeling. The explanation, in both cases, is probably that because of its strong binding to DNA, a high proportion of total tissue drug is DNA bound. The model predicted that Adriamycin levels in all tissues would fall in parallel, with a terminal half-life of about 16 h. For most of the tissues the Harris and Gross model gave a good fit to the experimental data, the chief exception being bone marrow, where the measured drug concentration at 24 and 48 h was severalfold higher than the predicted level and where it seemed to be dropping more slowly than in other tissues. However, despite its limitations, it was clear that the Harris and Gross model provided a wealth of useful information about Adriamycin pharmacokinetics and provided a degree of insight that would not have been attainable from a classical PK model. A version of this model was used by Chan et al. (1978) to predict Adriamycin tissue distribution in human cancer patients.

C. Some Other Flow Limited PB-PK Models

One of the first drugs modeled using the PB-PK approach was thiopental, which was studied by Bischoff and Dedrick in 1968 (see discussion by Bischoff, 1986). This was a relatively simple model, using four compartments — blood, viscera, lean tissue, and adipose. This early model thus represented an intermediate stage between a classical model and the more detailed physiological models, such as that for Adriamycin, discussed above. However, the model very successfully predicted blood levels of thiopental in dogs and humans. Thiopental is very lipid soluble, and the detailed description of an adipose tissue compartment was undoubtedly a key feature of this model. Another interesting feature of the Bischoff and Dedrick thiopental model was that it modeled tissue blood content explicitly. The model of Figure 2.1 assumes that arteries deliver drug to a tissue and veins remove it, but assumes that within the tissue drug is in homogeneous solution. The thiopental model, in contrast, assumes that tissues contain a pool of blood, whose drug concentration is in equilibrium with that of the extracellular fluid (Figure 2.2). If the rate constants for drug entry and exit between capillary and extracellular fluid are high, the distinction between drug inside capillaries and drug in tissue disappears, and the compartments collapse into the "lumped compartments" of Figure 2.1. Conversely, the distinction becomes more important for the transport-limited models discussed below.

Dedrick et al. (1972) constructed a model of cytarabine (AraC) pharmacokinetics that used a flow limited approach. In fact, cytarabine is actively transported, but the transport in most tissues appears to be so rapid that blood flow is still limiting for drug accumulation. Dedrick et al.'s cytarabine model was notable in that it incorporated *in vitro* data on drug metabolism, derived from human tissues. The seven-compartment model included blood, liver, kidneys, heart, "lean tissue" (muscle), gut, and bone marrow. Cytarabine is not lipid soluble, so no adipose compartment was necessary, and the compound is highly water soluble and negligibly protein bound, so that the plasma–tissue partition ratios were assumed to be one for all

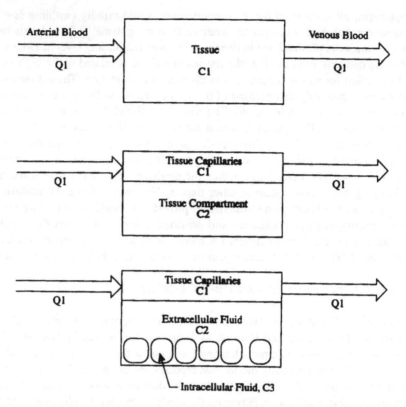

Figure 2.2 Different approaches to PB-PK modeling of the drug content of tissues. The "lumped compartment" approach (top) assumes that drug entry and exit between capillaries, extracellular fluid, and intracellular fluid is high. If we assume that tissues drug levels are in equilibrium with, but not necessarily identical with, extracellular fluid, the middle model is obtained. Finally, those drugs that require active transport, or where cellular accumulation occurs, may have intracellular drug levels that differ from extracellular fluid (bottom).

tissues. Cytarabine is deaminated to an inactive metabolite, arabinosyluracil (AraU), by cytidine deaminase, and this was assumed to take place in liver, heart, and kidney. The deamination reaction follows Michaelis–Menten kinetics, and the V_{max} and K_m values were measured in human tissue homogenates. Phosphorylation or other reactions of AraC metabolism were not modeled. Both AraC and AraU were cleared exclusively through the kidney. The model was designed to be consistent with data based upon experiments measuring total radioactivity following administration of radiolabeled drug, and it was used to predict plasma concentrations of AraC and AraU in man after i.v. injection of 1.2 and 86 mg/kg of the parent compound. In general, the predicted values agreed fairly well with experiment. Two conclusions that emerged from this study were, first, that the lean tissue compartment, which is relatively poorly perfused, acted as a reservoir, thus maintaining plasma levels for a relatively long time, and second, that because of the rapid deamination in some tissues, a large fraction of the drug may be deaminated on a single pass, making the observed kinetics strongly dependent upon blood flow rate.

In addition to humans, Dedrick et al. (1973) studied AraC pharmacokinetics in mice, dogs, and monkeys. Across the four species, the K_m of cytidine deaminase for AraC differed by more than 7-fold, and the V_{max} values varied over a 17-fold range, but use of the appropriate values gave a good fit to the data in every case. Across the four species the renal clearance varied according to an allometric (power law) relationship.

The Dedrick cytarabine model was further elaborated by Morrison et al. (1975) who added an additional biochemical factor, the phosphorylation of AraC to its active metabolite, araCTP, and dephosphorylation of araCTP back to AraC. Because the spleen is a major site of AraC phosphorylation, they added an additional compartment to represent this organ. A very important conclusion of the study of Morrison et al. was that araCTP stays around much longer in most tissues than the parent AraC, so that if pharmacodynamic conclusions are to be drawn, it is essential to model the phosphorylation step. The same research group (Lincoln et al., 1976) went on to study a model that combined this PB-PK model and intracellular enzyme kinetic model with a cytokinetic model of L1210 leukemia and used it to simulate treatment of L1210 mouse leukemia with AraC. This study suggested that the critical factors in determining response to AraC were deoxycytidine kinase activity, araCTP half-life, renal clearance of AraC, and cell kinetic parameters for cell proliferation and cell killing.

The work described in these four papers (Dedrick et al., 1972, 1973; Morrison et al., 1975; Lincoln et al., 1976) represents a milestone in the modeling of drug effects. These studies incorporated in vitro enzyme parameters for the first time into a PB-PK model and went on to combine this model with a cytokinetic model. The studies ranged across four species and have considerable experimental validation. The results emphasized the importance of cellular pharmacokinetics as a determinant of drug action, a conclusion relevant to all nucleotide drugs that are administered as an inactive nucleoside precursor and applicable to many other drug classes as well. The results of these studies have indirectly affected our thinking about nucleoside drugs and probably influenced the development of later pyrimidine nucleosides, such as gemcytabine. What is hard to understand is why this seminal work, conducted in the early to mid-1970s, has had so few successors. This 20-year-old work remains one of the most (perhaps still the most) comprehensive computer models of drug action. This is probably a reflection of the small numbers of pharmacologists qualified to do such elaborate modeling, coupled with a conviction that this model is too difficult and too expensive to be used as a standard tool in drug development. The author believes that developments in computer hardware and software and in analytical methodology have made it feasible to do modeling of this complexity as a standard part of drug development, and that doing so would provide insights that would amply justify the costs.

2-Amino-1,3,4-thiadiazole (ATDA) is an anticancer agent that requires intracellular activation to a ribonucleotide metabolite, which then acts as an antipurine agent. King and Dedrick (1979) created a five-compartment flow limited PB-PK model to describe the kinetics of ATDA in mice, dogs, and monkeys. The model consisted of blood, lean tissue, liver, kidney, and gut compartments, and the kidney was the primary site of elimination. Renal excretion was treated as first order, but metabolic activation was treated as saturable (except in mice, where it was modeled as first order).

King and Dedrick's ATDA model was used to simulate the concentrations of ATDA and its metabolite in blood and tissues, and to predict urinary excretion. The results were mixed; the model generally gave an accurate description of the concentration–time profile of ATDA in plasma of mice and dogs, though it tended to underestimate blood levels of the metabolite at early time points, which the authors suggested may be because metabolite transport in tissues is not rigorously flow limited, so that its volume of distribution could increase with time. The model tended to overpredict urinary excretion of both ATDA and metabolite in all three species.

The model was used to predict concentrations of ATDA in monkey plasma following i.v. injections of 0.3, 3, and 30 mg/kg. The model correctly predicted plasma concentrations of ATDA at 0.3 and 3 mg/kg, but overestimated blood levels at 30 mg/kg. The elimination half-life increased with increasing dose.

The authors studied interspecies correlations of ATDA metabolism and renal clearance by calculating allometric coefficients (see Chapter 3). The allometric coefficient for renal clearance was 0.70, close to the literature value of 0.77 for inulin. For metabolic clearance the coefficient was 1.29, indicating that metabolism makes a proportionately greater contribution in larger animals, an effect likely to be particularly marked in humans.

A flow limited PB-PK model of the oral hypoglycemic agent tolbutamide in rats was published by Sugita et al. (1982). This model is of particular interest as it was an early demonstration of the use of PB-PK modeling in the study of drug–drug interactions, in this case of the effect of sulfonamides on tolbutamide kinetics. The model included blood and 11 tissue compartments (liver, kidney, muscle, adipose, heart, lung, skin, pancreas, spleen, gut, and brain), and arterial and venous blood concentrations were modeled separately. The three sulfonamide antibacterial agents sulfamethoxazole, sulfadimethoxine, and sulfaphenazole were studied, in particular, their influence on protein binding of the structurally relate tolbutamide (which is normally >90% protein bound in plasma) and on its metabolism in liver. The tolbutamide tissue–plasma partition ratios (the R values of Equation 2.4) were determined in the presence and absence of the sulfa drugs. Experimentally, it was found that the sulfa drugs increased tissue levels of tolbutamide and slowed down its elimination. The model accurately predicted the concentration–time profiles of tolbutamide in most compartments, including the target organ, pancreas (though predicted levels of tolbutamide in the brain of non-sulfa-treated rats were low). The model also accurately predicted the free tolbutamide concentration in the rat pancreatic vein. This ability to model free drug concentration at its physiological site of action, and the accurate prediction of an important drug–drug interaction make this study an excellent example of the value of the PB-PK approach.

In addition to its applications in pharmacology, the physiological modeling approach has also been extensively used in toxicology to study the bodily disposition of toxic substances. An early example was the model of kinetics of the chlorinated hydrocarbon insecticide chlordecone (Kepone) in rats by Bungay et al. (1981). This model was unique in its highly detailed representation of the gastrointestinal tract, which was modeled as six main compartments (stomach, upper, middle and lower small intestine, cecum, and large intestine). Each of the three segments of small intestine was subdivided into five subcompartments. Within each segment, chlordecone absorption was described in terms of four parameters: the blood and tissue concentrations of chlordecone, the R-factor, and a parameter that combined a permeability coefficient with the surface area available for uptake. The lumen content flow rate and blood flow rates were determined experimentally. The model was used to simulate the concentration–time profiles of chlordecone after oral and i.v. doses. In addition to predicting chlordecone concentrations in the various compartments of the gastrointestinal tract, the model was used to investigate the ability of cholestyramine (a waxy, nondigestible substance) to absorb toxins that, like chlordecone, diffuse into the gut from the blood or toxins that are excreted in the bile.

D. Transport-Limited PB-PK Models

Chen and Gross (1979) constructed a PB-PK model of disposition of the anti-cancer agent, actinomycin-D, in dogs. This model was generally similar to the Adriamycin model described above, but modeled an additional compartment, the testes. When the assumption was made that drug uptake into the tissues was blood flow limited (as was assumed for all the models discussed above), an accurate description of actinomycin-D kinetics was obtained for gut, liver, and muscle. However, drug equilibrated into the testes more slowly than into other tissues and declined more slowly. For a flow-limited model, tissue drug levels may peak at different times for different tissues, but after they peak, the levels decline at the same rate in all tissues, i.e., as a series of parallel lines on a plot of log drug concentration vs. time (linear); Figure 2.3a illustrates this behavior. The concentration–time curves for the various tissues not being parallel on a semilog plot suggests that the flow-limited assumption is not valid. Chen and Gross (1979) then modeled drug uptake into the testes assuming that the net flux, r, was proportional to the concentration gradient across the cell membrane:

$$r = k \ (C_E - C_i) \qquad (2.8)$$

where C_E and C_i are the drug concentrations in extracellular and intracellular fluid, respectively, and k is a rate constant that for this particular simulation was set to 0.2 h^{-1}. Figure 2.3b shows the concentration–time curves for the testes as calculated by this transport-limited model, which gave a reasonably good fit to the experimental data.

E. The Enterohepatic Circulation

Tterlikkis et al. (1977) developed a six-compartment flow limited model of the pharmacokinetics of 6-mercaptopurine (6-MP) in rats. The compartments modeled were plasma, liver, small intestine, kidney, bone marrow, and spleen. 6-MP is a purine base analogue that is metabolized within cells to a nucleotide that binds tightly to its target enzyme (inosine 5′-phosphate dehydrogenase). This was handled in the model by a nonlinear tissue binding term. The model gave reasonable predictions of

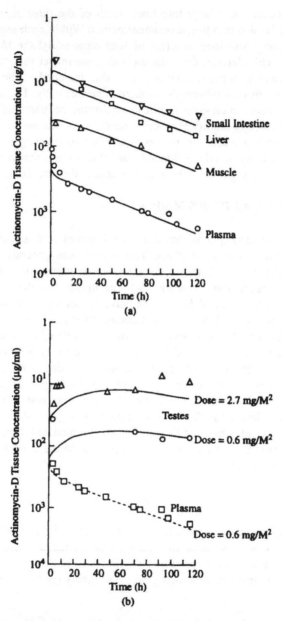

Figure 2.3 Tissue concentration profiles of actinomycin-D in dogs after 0.03-mg/kg i.v. (a) Flow rate-limited model; (b) transport-limited model. (From Chen and Gross, 1979. Reproduced with permission of Springer-Verlag.)

the 6-MP concentration–time curve for plasma, kidney, and spleen, but the fit for liver was not so good. In general, the model tended to overpredict tissue levels at early time points and underpredict at the later time points; probably because the complex metabolism of 6-MP was being seriously oversimplified. An important advance made by this particular model was that it was one of the earliest PB-PK models that attempted to describe the enterohepatic circulation. 6-MP is secreted by the liver into the bile, the bile is released into the small intestine, and part of the 6-MP is then reabsorbed into the blood. The effect of enterohepatic circulation is to prolong plasma levels of drug for longer than would be expected.

Another interesting aspect of this 6-MP study was that the authors were able to extrapolate the rat data to a model of 6-MP disposition in humans. The predicted human drug concentrations were compared with experimental data obtained for plasma, bone marrow, and spleen in four patients, giving excellent agreement.

F. Methotrexate Pharmacokinetics

The anticancer and antiarthritic agent methotrexate (MTX) is one of the most intensively studied of all drugs; it was one of the first tight-binding competitive inhibitors whose K_i was accurately known, one of the first drugs whose binding to target enzyme was visualized at the molecular level by X-ray crystallography, and the first drug to which resistance was shown to result from gene amplification. MTX is carried across cell membranes by a carrier protein whose physiological role is transport of various anionic molecules, including the natural folate cofactors. This reduced folate carrier (RFC) is essential for efficient transport of MTX (at very high plasma MTX concentrations, some may get into cells by passive diffusion). An accurate PB-PK model thus requires using a transport-limited approach. Bischoff et al. (1970, 1971) published the first PB-PK model for methotrexate; this model was subsequently modified by Zaharko et al. (1971) and by Dedrick et al. (1973). It was used by Lutz et al. (1975) to describe the kinetics of MTX in a spontaneous canine lymphosarcoma, and again by Zaharko et al. (1974) to study the relative toxicity of MTX in tissues of mice bearing Lewis lung carcinoma. Dedrick et al. (1978) added an additional compartment, the peritoneal cavity, to the model in order to use it to study the intraperitoneal (IP) "belly bath" route of administration of high dose MTX for treatment of ovarian cancer. The model showed that after IP administration local concentrations of MTX were retained in the peritoneum that were much higher than the plasma levels, thus conferring a therapeutic advantage for treatment of tumors that are confined to the peritoneum. This model included enterohepatic cycling, which is important because (as for 6-MP) a significant amount of MTX is excreted in the bile and is then efficiently resorbed from the small intestine.

Another feature of MTX that has a strong influence on MTX pharmacokinetics is its very tight ($\sim$5 pM) binding to its target enzyme, dihydrofolate reductase (DHFR). Thus, tissues that have high levels of DHFR (GI mucosa, bone marrow, spleen, and tumor) will accumulate higher total MTX than nonproliferating tissues such as liver or muscle. As a first approximation it was assumed by these early models that for a pharmacodynamic effect (e.g., cytotoxicity) to be seen, the level of MTX inside the target cell has to be sufficient to titrate the DHFR and then accumulate free drug. This approximation, though crude, is fairly effective in predicting response and is a consequence of MTX being a very tight-binding inhibitor of a non-rate-limiting enzyme (though it is not, in fact, an irreversible inhibitor, as sometimes stated in the earlier literature). The use of models that combine biochemical information with this PB-PK model makes it possible to study the effects of, for example, mutations in tumor cells that decrease the drug transport activity, mutations that increase the Ki for binding of MTX to DHFR, or drug resistance resulting from DHFR overproduction.

III. THE BLOOD–BRAIN BARRIER

Dedrick et al. (1975) reported measurements of MTX in plasma and cerebrospinal fluid (CSF) of a patient who had received a 24-h i.v. infusion. At the end of the infusion the concentration of MTX in CSF was about 5% of the plasma level. Following the infusion, the CSF concentration, however, declined more slowly than the plasma level. This is a typical example of the functioning of the blood–brain barrier (BBB). The BBB is the result of very tight intracellular junctions between the endothelial cells of brain capillaries, which acts to control the biochemical environment of the brain and to protect it from potentially neurotoxic substances. The BBB is a two-way barrier, which is why the CSF concentration of MTX declines so slowly. Dedrick et al. (1975) proposed a compartmental model to describe drug transfer between plasma and CSF. This model is illustrated in Figure 2.4. The mass balance equation used for this model was

$$V_B \cdot dC_{CSF}/dt = r + (PA)(C_B - C_{CSF}) - Q_{CSF} \cdot C_{CSF} \qquad (2.9)$$

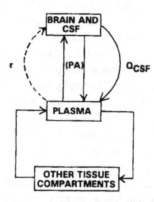

Figure 2.4 Compartmental model of drug transfer between plasma and CSF. PA = permeability × area for passive transport; r = rate of active transport; Q_{CSF} = CSF flow rate. (Adapted from Dedrick et al., 1978.)

It contains a passive transfer component consisting of a permeability factor (PA), where P is a coefficient related to the the physical properties of the drug and A is the area available for exchange; r is the rate of active transport, and Q_{CSF} is the rate of removal of drug from brain by bulk flow of CSF. This kind of model is termed a hybrid PK model. Hybrid models combine features of classical compartmental models and of PB-PK models; hybrid models generally focus on a single organ (the brain, in the present case). They are physiological in the sense that the compartment being modeled has a defined anatomical meaning, but the drug concentrations entering the organ are expressed as multiexponential equations derived by a nonlinear regression fit to experimental data, as for a classical PK model. Dedrick et al. (1975) went on to model the direct administration of MTX in CSF by intrathecal injection and showed that this resulted in much higher brain levels of drug. Plasma MTX peaked at a much lower level than if the same dose had been administered IV, but lasted much longer (which could increase systemic toxicity).

Gallo (1991) reported a hybrid PK model characterizing the disposition of the anti-AIDS drug azidothymidine (AZT), and of its metabolite AZddU, in the mouse brain. His model was based upon an earlier model written by Spector et al. (1977) to describe concentrations of mannitol and ascorbic acid in brains of rabbits. This model decribed transfer of these substances between plasma, CSF, and intracellular fluid of brain and allowed the transfer to be first order or nonlinear (Michaelis–Menten) processes. Distribution of AZT to brain is a subject of therapeutic importance, since human immunodeficiency virus may enter the brain where it may cause AIDS-related dementia. Gallo treated the passage of AZT across the BBB as a Michaelis–Menten process, and tranfer of AZT between CSF and brain intracellular fluid as passive. His predicted brain concentrations of AZT in rabbits showed good agreement with the experimental data.

A. Prediction of Blood–Brain Barrier Penetration

There is considerable interest among drug developers in developing computational techniques to predict ability of novel compounds to penetrate the BBB. If we are developing drugs intended to act within the CNS, then BBB penetration is a prerequisite; for most drugs with a site of action outside the CNS, however, penetration of the BBB is a liability, since it confers no advantage and opens up the possibility of neurotoxicity. For anticancer agents the picture is more complex; we would like to have some antitumor drugs that penetrate the barrier, in order to treat brain tumors as well as CNS metastases of tumors that have arisen at other sites. In large tumors (which generally have aberrant capillary morphology) the BBB breaks down, so drug delivery is presumably not a problem. However, this is not the case for micrometastases, which have not yet established their own blood supply. Since, in general, small metastases are more curable than established tumors, the inability of anticancer drugs to penetrate the barrier in this context (e.g., in adjuvant chemotherapy) is a serious limitation.

To predict passage of compounds across the barrier, it helps to understand its nature. Most capillaries are relatively leaky; they keep blood cells inside, but allow passage of small molecules (and even protein molecules) to pass through the cracks between the endothelial cells. This is why, in general, the assumption that drug concentrations in the extracellular fluid in most tissues are in equilibrium with blood levels is a good approximation. In the brain, however, the endothelial cells have tight (waterproof) junctions between adjacent cells. Thus, to enter the brain, substances in the blood plasma must diffuse through the endothelial cells, rather than between them. In order to partition from the aqueous environment of the plasma into the nonpolar environment of the endothelial cell membrane, a drug must be relatively lipophilic. Medicinal chemists generally describe lipophilicity in terms of log P, i.e., the log of the n-octanol–water partition coefficient; a log P of zero implies that the molecule is equally soluble in n-octanol and water, negative log P indicates greater water solubility, and positive log P greater lipid solubility. Log P has been determined traditionally by direct partitioning and more recently by HPLC (see Seelig et al. [1994] for references). Goldstein and Betz (1986) show a relationship between log P and BBB penetration for a range of drugs: for log P more polar than -2, penetration is very low; median BBB transport was seen at about log $P = -0.8$; and for log $P > 0$, BBB penetration was maximal. In practice, very lipophilic drugs sometimes penetrate worse than moderately lipophilic compounds; this may be because the drug must not only penetrate into the endothelial cell membrane, but once in the membrane must be able to equilibrate back into the aqueous phase on the other side of the membrane. Perhaps extremely greasy drugs simply enter the membrane and stay there.

The relationship between log P and BBB transit applies only to those drugs that must cross by passive diffusion. Brain endothelial cells appear to be richly supplied with an array of active transport carriers, possessing a range of chemical selectivity, that pump nutrients across the endothelium, in some cases against a concentration gradient, which requires the expenditure of energy. The BBB thus exercises strict control over which compounds enter the brain from the blood. This transport may be asymmetric; for example, glycine can be pumped out of the brain but not into it. There is a net flux of potassium ions out of the brain and of sodium ions into it (Pardridge, 1988).

The rule that log P predicts for BBB permeability is a first approximation only, and, even neglecting those compounds that may be transported by one of the specific carrier mechanisms, there are many exceptions in both the high- and the low-lipophilicity range. Seelig et al. (1994) have proposed that (absent active transport) three properties of a drug molecule will determine its ability to cross the BBB: (1) the number of lipophilic groups, (2) the number of charged groups and their extent of ionization, and (3) the molecular size. They point out that these same three factors also determine the surface activity of a molecule, and that their combined action can be determined in a single experiment by measuring the Gibbs adsorption isotherm, which is done by examining the effect of the compound on surface tension in a water–methanol solution. Seelig et al. observed that when surface pressure, π (which is the difference between surface tension of pure buffer and the surface tension of the buffer solution containing the drug), was plotted as a function of concentration, at very low drug concentrations no surface effect could be seen. As the drug concentration was increased, a concentration, C_o, was reached at which surface pressure began to increase with concentration, at first slowly and then reaching a region where the plot of π against log C was linear. Finally, a concentration was reached above which no further increase in π was obtained (the critical micelle concentration, or CMC). If CMC was now plotted against the log of C_o, the 25 compounds fell into three groups: the first region corresponded to very hydrophobic compounds with poor transport across the BBB, the second region to moderately hydrophobic compounds with good penetration into the CNS, and the third region to more hydrophilic compounds that only penetrated the BBB at very high concentrations. The correlation was best when the CMC–log C_o plot was done at pH 8.0 for cationic compounds and at pH 6.8 for anionic compounds. The rationale for selection of these pH values was as follows: blood plasma has a pH of 7.4; however, the penetration of a charged molecule into a nonpolar environment is energetically unfavorable, and the molecule will reduce its charge, provided the pK_a of the charged residue is about 7 to 8. For tertiary amines the apparent pK_a shift is about +0.6, and for the anionic compound thiopental, about –0.6. The predictive power of the corrected graph of CMC–log C_o was much greater than that of the n-octanol–water partition coefficient.

Clearly, the ability to predict passage of novel compounds across the BBB would be of great value. Such predictions, used in conjunction with PB-PK models, should make possible more accurate predictions of efficacy for CNS-active agents and of neurotoxicity for compounds that act outside the brain.

IV. DESIGN OF PB-PK MODELS

A. How Many Compartments Are Necessary and Which Should They Be?

Obviously, the choice of compartments is influenced by the particular drug under study and its target organs for efficacy and toxicity. However, as pointed out by Bischoff (1986) the design of PB-PK models imposes some obvious constraints. Bischoff listed three necessary compartments or groups of compartments for a minimal model: (1) the efficacy and toxicity target organs, (2) the organ or tissue to which most of the drug distributes (which, depending upon the physical properties of the drug will often be muscle or adipose tissue), and (3) the major site or sites of drug elimination, usually kidney or liver. If enterohepatic circulation is likely to be a major determinant of overall pharmacokinetics, then the gut and portal circulation must be modeled. Sometimes organs and tissues may be combined into "body regions". For anticancer drugs, a separate tumor compartment will often be necessary. For anesthetics, whose rate of action is very fast, the usual assumption that the entire blood volume is well mixed and the plasma drug concentration is uniform may no longer be true, so that the models become much more complex. Bischoff (1986) discusses a model in which the blood is modeled as 26 distinct compartments.

B. Modeling Drug Levels in the Lungs

If we wish to predict drug levels in the lungs, these have to be modeled a little differently from other tissues; because of the double circulation system of higher vertebrates, the entire cardiac output passes through the lungs. This characteristic of the lungs is reflected in the position of the lung compartment in Figure 2.1. It follows from this position that if a drug is injected intravenously, the first capillary bed it will encounter is that of the lungs. This can have some practical consequences; e.g., if a particular drug formulation has poor solubility and has a tendency to precipitate after i.v. injection, the particles will be deposited in the lungs. If this happened in the course of an ADME study, it could suggest a spuriously high affinity of the test compound for lung tissue.

The mass balance equation for the lung is

$$dC_{lu}/dt = \sum_{i=1}^{i=n}(Q_i \cdot C_i/R_i) - Q_{lu} \cdot C_{lu}/R_{lu} \tag{2.10}$$

where the subscript lu denotes lung and the summation term applies to all other compartments (except blood). Note that $Q_{lu} = Q_B$.

C. Causes of Nonlinear Disposition

First-order elimination is usually an accurate description of renal elimination for those drugs that enter the renal tubules through glomerular filtration. If compounds are actively secreted into the urine, this process will be saturable and will have to be described by a Michaelis–Menten equation. The same applies to secretion from the liver into the bile, which is always an active process. Reactions of drug metabolism, being enzymatically catalyzed, are also saturable and will also follow Michaelis–Menten (or more complex) kinetics. It is sometimes stated that ethanol is unique in being eliminated by zero-order kinetics; Welling (1986) points out that the quantities of alcohol consumed in alcoholic drinks are large in comparison with the usual doses of most other drugs and that the alcohol dehydrogenase reaction is saturated under these conditions, giving apparent zero-order kinetics. Frequently a drug is metabolized by several different pathways. If these are operating at different regions on their saturation curves, it may be convenient to model the metabolism by the sum of saturable and nonsaturable terms.

Estimates of K_m and V_{max} are sometimes calculated from *in vivo* experiments. It may often be easier and more accurate to determine these values *in vitro*, using tissue homogenates or purified enzymes. The *in vivo* and *in vitro* results may not always agree; it is important to remember that the *in vivo* data are apparent kinetic parameters, which will reflect the presence of any endogenous inhibitors or competing substrates. Of course, such apparent values are the pharmacologically relevant ones.

Nonlinearity is sometimes seen in the tissue–plasma partition coefficients, which have been treated so far as constants, but which in some cases show concentration dependence. One explanation of this is tight binding of the drug to a specific enzyme or receptor in the tissue. In such cases the drug concentration in a tissue is the sum of a tightly bound component that will depend on the tissue concentration of the enzyme or receptor and a free (and/or weakly bound) component.

D. Prediction of Oral Drug Absorption

Oral administration in a PB-PK model can be simulated by including a gut lumen compartment (as in Figure 2.1) and modeling rate of drug uptake into the gut epithelium compartment (Q_{20} in Figure 2.1) as well as bulk transport of gut lumen contents out of the system (Q_{19}). The absorption of drug from gut lumen into epithelium may be passive (first order) or active (saturable). Modeling oral absorption requires an estimate of oral bioavailability, which can be measured *in vivo* by comparing AUC values after i.v. and oral administration (this may not be possible if an i.v. formulation for the drug is not available, which is often the case). Alternatively, permeability factors may be measured *in vitro* or estimated from the physical properties of the compound, as discussed in Chapter 4. These factors may then be used in a PB-PK model. For drugs in which absorption occurs at different parts of the GI tract (e.g., stomach and small intestine) it may be necessary to add additional compartments to the model.

V. MODEL FITTING

The classical approach to parameter estimation for PK models is to do a nonlinear regression fit to the experimental data, in which estimates are generated for each of the system parameters. This would only work for the simplest PB-PK models. For a typical PB-PK model, which is described by dozens of parameters, the classical approach would simply break down. A discussion of parameter estimation for PB-PK models is given in Gallo (1991). A common approach is to use a combination of literature data (for blood flow rates, in particular, and usually for tissue volumes) with experimentally fitted data (e.g., for tissue to plasma partition ratios). Tables of blood flow rates and tissue volumes for several different species are given in Chapter 3; these are based upon allometric equations. An invaluable compilation of experimental values for six species (mouse, rat, rabbit, dog, monkey, and human) is given by Davies and Morris (1993). This reference includes tables for organ weights, organ and body fluid volumes, blood flows (and urine and bile flow rates), information of intestinal transit times, pH values, as well as information on heart rate, oxygen consumption, and other useful data.

As mentioned above, a common approach is to use literature values for blood flows and organ volumes, and then estimate the tissue–plasma partition ratios from the experimental time-course of tissue drug levels, using any standard nonlinear regression program (e.g., SIMUSOLV, or STELLA II — see the Appendix). For an example of the use of SIMUSOLV in PB-PK modeling, see Frederick et al. (1992). Mitchell and Gauthier Associates, Inc., the U.S. and Canadian distributors of SIMU-SOLV, distribute a bibliography of its pharmacological applications.

Ideally, levels of drug in various tissues should be corrected for the blood content of the tissue. Values for 12 commonly modeled tissues are listed in Ebling et al. (1994) based upon data obtained by Everett et al. (1956) in liquid nitrogen-frozen samples.

A shortcut for getting tissue–plasma partition ratios from *in vivo* data is to use the results of the ADME study in rats, since this is part of every drug's standard preclinical development. Usually, this study follows a dozen or more different tissues and has multiple time points, usually out to 24 or 48 h. Just comparing tissue–plasma ratios at a particular time is the easiest first approximation, though, of course, there is no guarantee that the tissue is in equilibrium, especially at early time points. A more rigorous analysis of the data can be obtained by doing a nonlinear regression fit to the entire time-course, as described above. A limitation of using ADME data, instead of a custom-designed tissue distribution study, is that the ADME data are often only obtained at a single dose level.

An alternative approach is to measure the tissue–plasma partition ratios *in vitro*, using tissue slices or homogenates. This has the disadvantage that it is more work, but the advantage that it can give very accurate estimates of true equilibrium ratios. If human tissue is available, then measured values can be used to create a direct PB-PK model of drug distribution in humans. Lin et al. (1982) measured tissue–plasma partition coefficients for ethoxybenzamide *in vitro* for nine tissues (using tissue homogenates) and *in vivo* by two routes (following bolus injection and following constant rate infusion). The values obtained by these three methods were in good agreement.

Ebling et al. (1994) argue that the parameter-fitting problem for a PB-PK model of any complexity is so formidable that there is a tendency to approach it as a simulation exercise and fit parameters by eye rather than by objective optimization. They have proposed a new approach in which drug disposition is described in each tissue separately, using constrained numerical deconvolution. The time-course of drug concentration in each tissue is described as the convolution of an input function with a unit disposition function. In effect, separate parameters are derived for each tissue, and their estimates are independent of the estimates for the other tissues. The entire body is then described by assembling the individual submodels. Ebling et al. (1994) illustrate this process with a 15-tissue model (total of 27 compartments) of disposition of the anesthetic thiopental in rats. This model is illustrated in Figure 2.5. The completed PB-PK model gave excellent agreement with the measured data.

When modeling drugs where tissue availability is transport limited, the parameter estimation problem is made even more complex by the need to estimate transport parameters (either a first order rate constant or, for saturable transport, V_{max} and K_m) for drug uptake. This must be done for each tissue, greatly increasing the total number of parameters. As for flow limited models, sometimes quite successful attempts have been made to fit all the parameters by fitting the whole data set by nonlinear regression techniques, though this is very difficult. Gallo (1991) discusses methods of fitting transport parameters to *in vivo* data. He criticizes *in vitro* approaches on the grounds that parameters obtained *in vitro* with a single cell type may not apply to other cell types. If a single carrier model functions in all tissues, it is likely that the K_m value will be the same in all tissues, but that V_{max} may differ from one tissue to the next, since some tissues will have higher expression levels than others. One approach is thus to determine K_m with a single cell type and fit V_{max} for each tissue from the *in vivo* data. In the author's opinion, a better way (though more time-consuming) would be to determine K_m *in vitro* from detailed studies with a single cell type, and then determine V_{max} *in vitro* for each tissue type, using tissue slices or (better) isolated cell suspensions. Where K_m is high relative to achievable plasma drug levels, the uptake may be approximated as an apparent first order process, which simplifies the parameter estimation. On the other hand, the process may be a good deal more complex. For the anticancer drug lometrexol, entry into cells is by either a high-affinity, low-capacity folic acid-binding protein (FABP) or by a low-affinity, high-capacity reduced folate carrier (RFC). Some tissues express one of these, some the other, and some express both, in various ratios. An understanding of drug transport, and the ability to model it, is crucial to an understanding

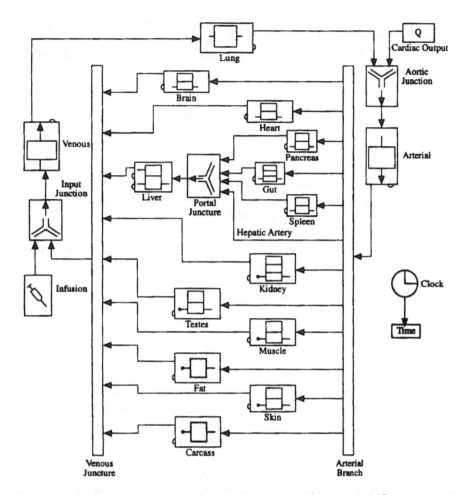

Figure 2.5 A PB-PK model for thiopental that incorporates 15 tissues and 27 compartments. Tissue symbols indicate the number of compartments within each tissue. (From Ebling et al., 1994. Reproduced with the permission of Plenum Press.)

of drug efficacy and toxicity for drugs of this class. In the case of methotrexate, drug efflux may be through the same carrier responsible for uptake (the RFC again), but efflux may also utilize two other carrier systems. When modeling drugs whose transport is very complex, the decision of how much detail to include in a model can be a difficult judgment call. It is hard to generalize, but *in vitro* studies will almost always be necessary.

Drug metabolism parameters present some similar problems to drug transport, though the problem will be easier if metabolism is confined to a single tissue (often liver). In this case it may be quite feasible to calculate the metabolism parameters from *in vivo* data, particularly if metabolism is apparent first order. Often, an HPLC assay for the drug under study may also reveal the metabolite or metabolites, whose rate of appearance in tissue or plasma may then be quantitated in detail. However, studies with liver slices, isolated hepatocytes, or liver microsomes (the latter particularly when metabolism is known to be via the cytochrome P450 system) can provide valuable kinetic data on metabolite formation, and this data may be readily incorported into a PB-PK model. *In vitro* techniques of measuring drug metabolism are discussed in Chapter 4.

VI. OTHER EXAMPLES OF PHYSIOLOGICAL PHARMACOKINETIC MODELS

The thesis of the present work, as repeatedly emphasized, is that to improve the success rate of developmental drugs, we need more detailed information on drug delivery and drug distribution, and for this reason the PB-PK approach is likely to become an increasingly important part of preclinical drug development. The current chapter describes PB-PK models for the anticancer agents Adriamycin, actinomycin-D, methotrexate, 6-mercaptopurine, 2-amino-1,3,4-thiadiazole and cytarabine, the antiviral azidothymidine, the anesthetic thiopental, the insecticide kepone, and the hypoglycemic agent tolbutamide. These ten examples have illustrated many of the principles of PB-PK modeling. However, it is worth briefly discussing a few further examples.

King and Dedrick (1981) described a six-compartment, flow-limited model for the antileukemic drug pentostatin (2'-deoxycoformycin). This model included a tumor compartment, which made it possible to compare the PK of pentostatin in normal and leukemic mice, and some differences were, in fact, seen. Intracellular binding was treated as the sum of linear nonspecific binding plus strong saturable binding to adenosine deaminase. Another interesting characteristic of the pharmacokinetics of pentostatin is that the kidney clearance exceeds the glomerular filtration rate, indicating that pentostatin must be actively secreted; it was modeled as a linear process involving both filtration and secretion.

Farris et al. (1988) described a seven-compartment, flow-limited model for the antineoplastic drug cisplatin. The model has been applied to several species — rat, rabbit, dog, and human. A notable feature of this model was that it simultaneously tracked three drug species — cisplatin, fixed metabolite (i.e., macromolecule-bound platinum adduct, formed from cisplatin), and mobile metabolite; the mobile metabolite is formed from both parent cisplatin and from the fixed metabolite. In fact, the active form of cisplatin is the aquated species, and the mobile metabolite is actually a lumped species that represents metabolites of both parent cisplatin and the aquated derivative. The model gave a good description of cisplatin kinetics in several species and also gave good predictions of the PK properties of cisplatin analogues. For example, because cyclobutanedicarboxylic acid (which replaces the chloride ions of cisplatin in carboplatin) is a less reactive leaving group than chloride, carboplatin forms macromolecular metabolites about ten times more slowly than cisplatin. The model correctly predicted that urinary excretion of carboplatin should be more extensive than that of cisplatin and that (at equal doses) carboplatin should give lower tissue platinum concentrations.

A final example from the anticancer field is the study by Guaitani et al. (1991) on the nitrosoureas carmustine (BCNU) and fotemustine (Servier 10036) in rats.

This study only modeled three compartments (plasma, liver, and tumor). Two different tumor lines were studied, a nitrosourea-sensitive line of the Walker-256 carcinoma and a nitrosourea-resistant variant of the same tumor. The distribution volume of both drugs in plasma was greater in tumor-bearing rats than in controls. Fotemustine was cleared two to five times more slowly from tumor tissue than was BCNU. The AUC of fotemustine in the nitrosourea-resistant tumor was about sixfold lower than in the sensitive tumor, suggesting that the resistant line had an increased rate of drug breakdown.

Björkman et al. (1994) have described a detailed 14-compartment model of distribution of the opioid analgesics fentanyl and alfentanil in rats and humans. The approach proceeded by the two-stage method (as described above for the work by Ebling et al. (1994). The Michaelis–Menten parameters for the hepatic metabolism of alfentanil were determined by iterative optimization of the entire model.

An excellent example from the cardiovascular field is the model of the antiarrhythmic agent, amiodarone, discussed by Yasuda et al. (1993). The PK of amiodarone have generally been considered to follow a three-compartment classical model, with slow equilibration into adipose tissue. Oral bioavailability and elimination half-life vary widely between individuals. Efficacy after oral dosing requires several days to become apparent, perhaps because of poor bioavailability, slow equilibration to heart, and accumulation of an active metabolite, desethylamiodarone. This complex pharmacokinetic behavior makes amiodarone an example of where a PB-PK model could give useful insight into what may be going on in the body. Yasuda et al. (1993) show measured and modeled levels in humans of desethylamiodarone in seven compartments (blood, fat, liver, lung, heart, muscle, and kidney), following a 400-mg i.v. dose of amiodarone, with excellent agreement.

The pharmacokinetics of the immunosuppressive SDZ IMM 125 (a cyclosporin analogue) has been described in a complex PB-PK model developed by Kawai et al. (1994). These authors found that a flow-limited model did not give an adequate description of the kinetics of SDZ IMM 125. They developed a 16-compartment model (14 organs, plasma, and blood cells) in which each organ or tissue contained five subcompartments: plasma, blood cells, interstitial fluid (in instantaneous equilibrium with plasma), intracellular fluid, and a bound, slowly interacting intracellular fraction. Binding of drug to three different plasma proteins (albumin, low-density lipoprotein, and high-density lipoprotein) was modeled, and the same proteins were assumed to be present, though at different concentrations, in interstitial fluid. Transfer of drug between interstitial fluid and intracellular fluid (and between plasma and blood cells) was calculated as a product of surface area and a permeability factor. Equilibration of SDZ IMM 125 between plasma and blood cells was quite slow, though a high level accumulated in blood cells. This model was developed in rats, but scaled to dogs and humans. Parameters for blood cell uptake were measured *in vitro*. The different rates of hepatic metabolism in different species were treated as proportional to the known ratios of cytochrome P450 type 3A-catalyzed cyclosporin-A oxidation. This difference of metabolic clearance rate between species did not follow an allometric relationship. The model of Kawai et al. probably represents one of the most complex PB-PK models yet constructed that still gives good agreement with experimental data.

The final example is another situation where a requirement for detailed pharmacokinetic information makes PB-PK modeling desirable: distribution of macromolecules, where delivery into tissues is presumably more difficult than for small molecules. Mariani et al. (1991) discuss the example of a monoclonal antibody radioimaging agent, radioiodinated MoAb KC4. These authors modeled the distribution of this antibody between plasma, liver, spleen, and bone marrow, and its appearance in urine. The liver and spleen curves were not a very good fit, illustrating the difficulty and uncertainties of modeling macromolecule kinetics. In a study discussed in the same paper by Mariani et al. (1991) of kinetics of another antibody, [111]Indium-labeled C110, the authors show that treating liver as two separate compartments (one representing immunoglobulin catabolism, and the other, nonspecific binding in the reticuloendothelial system) improved the fit. In this model, tumor tissue was also treated as two compartments. Clearly since drug delivery presents particular difficulties for macromolecules, there is a need for further PB-PK modeling studies of macromolecules, which should assist in the rational development of these novel therapeutic and diagnostic agents.

VII. THE PHYSIOPLEX MODEL

In order to make PB-PK modeling more accessible to pharmacologists who are not experienced computer programmers, a user-friendly, general-purpose PB-PK model has been written for nonspecialist users. This program, PHYSIOPLEX (see Appendix), comes with files containing values of physiological parameters for seven species (mouse, rat, guinea pig, rabbit, dog, monkey, and human). These preset values can be modified as necessary. The general topology of the model is shown in Figure 2.1, except that three additional compartments have been included (spleen, heart, and "remainder"). Any compartments that is not necessary to model or for which no data are available may be "switched off" by setting their blood flow to zero. As input for particular drugs, the program asks for molecular weight, tissue–plasma partition ratios for 12 tissue compartments, and first order rate constants for renal and hepatic clearance. Drug administration may be modeled following i.v. bolus, i.v. infusion (of any required duration), intrathecal infusion, or oral administration. The PHYSIOPLEX program is based upon an earlier, published program, PHYSIO (Jackson, 1992), but differs in three respects: (1) It contains four additional compartments (lung, spleen, heart, and remainder). (2) It provides automatic scaling across species, i.e., if a model is created using rat data, simply loading a different species file will give a first approximation of how that drug may be expected to behave in, for example, dog or human. This feature will be discussed further in Chapter 3. (3) The PHYSIOPLEX program is able to model the disposition of drug metabolites, a feature that is discussed below. The PHYSIOPLEX program is purely a simulation module — it cannot be used for data fitting, which must be done by one of the methods described above.

As an initial example of the use of the program, we will emulate the classical Adriamycin study of Harris and Gross (1975) that was discussed above. For demonstration purposes, the parameter values for four anticancer drugs (carmustine,

Table 2.1 PHYSIOPLEX: Physiological Pharmacokinetic Model Modeling Adriamycin in Rabbit

		3-mg/kg i.v. Bolus			
Time	Plasma	Muscle	Adipose	Tumor	Brain
0	127,655.2	0	0	0	0
4.00	157.0	2,683.3	1,686.4	10,594.7	1,464.0
8.00	93.7	2,796.1	964.7	5,606.7	775.0
12.00	65.8	2,576.3	618.6	3,717.1	515.0
16.00	50.3	2,268.5	445.0	2,761.8	383.2
20.00	40.3	1,957.4	344.6	2,179.6	302.6
24.00	33.1	1,672.4	278.0	1,775.7	246.6

Alpha-phase half-life = 0.12
Beta-phase half-life = 2.53
Gamma-phase half-life = 14.5

Time	Kidney	Bone marrow	Liver	GI mucosa	Gut lumen
4.00	87,400.7	7,737.7	5,508.7	7,879.1	8,020.8
8.00	49,626.2	4,812.3	2,952.7	4,253.6	8,247.8
12.00	31,834.2	3,119.3	1,973.8	2,854.1	6,897.0
16.00	22,942.1	2,211.8	1,470.9	2,129.6	5,482.2
20.00	17,791.6	1,687.3	1,161.1	1,681.1	4,317.2
24.00	14,365.0	1,348.0	945.2	1,367.9	3,420.1

Time	Lung	Spleen	Heart	Remainder
4.00	21,387.8	65,863.0	7,164.6	1,877.6
8.00	12,772.9	59,758.0	4,136.1	1,032.0
12.00	8,970.8	48,614.5	2,864.0	703.9
16.00	6,860.1	38,433.7	2,173.2	530.1
20.00	5,495.7	30,325.9	1,733.8	421.2
24.00	4,511.8	24,127.6	1,420.4	344.3

Drug eliminated in urine = 3.3%
Drug eliminated in feces = 11.7%
Drug metabolized = 60.9%
Renal clearance = 5.93 ml/h/g
Plasma AUC = 2.86 μMh
AUC in tumor = 145.4 μMh

Adriamycin, cytarabine, and methotrexate) are included with the program. For other drugs, the user is required to enter the parameter values, which are thereafter stored in the system's data base. The drug parameter values, like the species parameter values, may be readily modified from within the program. By loading the Adriamycin data, and the rabbit parameter file, and entering a dose of 3 mg/kg (i.v. bolus) the output shown in Table 2.1 was obtained. Drug levels in plasma and in 13 tissue compartments are tabulated as a function of time. Note that although, for reasons of space, the table only shows drug levels at 4-h intervals up to 24 h, the program can give output at any requested time interval and for any required total time. The program also generates estimates of alpha-, beta-, and gamma-phase half-lives in plasma and estimates the renal clearance, as well as the amounts of drug metabolized in liver during the period of the simulation and proportions of administered drug eliminated in urine and feces during the period of the simulation. Finally, the program prints out area under the curve (AUC) for plasma and for any selected tissue compartment.

As shown by the Harris and Gross data, Adriamycin levels in all tissues greatly exceed the plasma concentration, with particularly high levels in kidney and lung. The predicted values for plasma and for three tissues are plotted in Figure 2.6. Note the early, high peak level of drug in the heart and the much slower accumulation of drug in muscle. These predictions are in close agreement with those of the original Harris and Gross model.

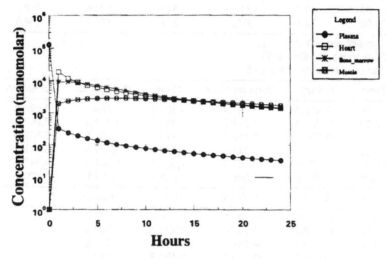

Figure 2.6   Adriamycin levels modeled using the PHYSIOPLEX program emulating the model of Harris and Gross (1975). Predicted levels are shown for plasma and for three of the nine tissues simulated.

It has been suggested that Adriamycin cardiotoxicity may be related to high peak plasma levels following i.v. bolus or rapid infusion treatment and that a more prolonged infusion of the same total dose may avoid this toxicity without compromising antitumor activity. PB-PK modeling provides an approach to predicting the differences in cardiac drug levels following infusions of different duration. An example of the use of the PHYSIOPLEX model to compare different kinds of drug administration is shown in Table 2.2. This shows the early time points only (out to 2.4 h for a 3 mg/kg i.v. bolus of Adriamycin in rabbit, compared with the same total dose administered as a 2-h i.v. infusion). As expected, the infusion avoids the very high initial plasma level. The plasma concentration rises steadily, peaks at 2 h, and then begins to drop with an initial half-time of about 0.3 h (this rate of decline subsequently slows down to a half-time of about 14 h). The peak drug level in heart following the infusion was about 50% of that following the i.v. bolus, while the peak tumor level was 85% of the level after an i.v. bolus. The AUC values in plasma and in tumor tissue measured over the first 24 h were within 1.5% for the two modes of delivery. The AUC values for heart were also similar for the two treatments, though, as stated above, there was a large difference in the *peak* values. This ability

Table 2.2 PHYSIOPLEX: Physiological Pharmacokinetic
Model Modeling Adriamycin in Rabbit

	3-mg/kg I.V. Bolus				
Time	Plasma	Muscle	Adipose	Tumor	Heart
0	127,655.2	0	0	0	0
0.20	329.2	1,406.5	2,152.3	22,662.2	41,431.6
0.40	364.9	1,536.0	2,199.6	21,975.4	31,788.3
0.60	348.4	1,663.4	2,243.1	21,397.4	25,684.8
0.80	326.3	1,779.8	2,268.8	20,752.9	21,483.8
1.00	305.9	1,885.2	2,277.7	20,045.5	18,484.4
1.20	287.8	1,980.2	2,272.9	19,303.9	16,291.9
1.40	271.7	2,066.9	2,257.0	18,541.9	14,628.2
1.60	257.3	2,145.7	2,232.2	17,779.4	13,337.8
1.80	244.3	2,217.3	2,200.5	17,028.0	12,309.0
2.00	232.6	2,282.5	2,163.3	16,295.9	11,467.1
2.20	222.0	2,341.9	2,122.0	15,588.7	10,761.6
2.40	212.3	2,396.0	2,077.5	14,910.1	10,158.0

	3-mg/kg 2-h I.V. Infusion				
Time	Plasma	Muscle	Adipose	Tumor	Heart
0	0	0	0	0	0
0.20	256.1	65.6	105.0	1,146.2	2,491.5
0.40	335.5	193.6	300.2	3,198.1	6,248.4
0.60	379.9	343.3	515.5	5,359.1	9,482.5
0.80	416.5	507.6	738.2	7,496.0	12,143.1
1.00	449.7	683.9	964.3	9,575.7	14,364.3
1.20	480.5	869.9	1,190.5	11,577.2	16,256.5
1.40	509.5	1,066.7	1,417.6	13,515.6	17,923.7
1.60	536.9	1,272.2	1,463.3	15,378.4	19,414.5
1.80	562.8	1,485.7	1,866.6	17,165.3	20,769.3
2.00	587.4	1,706.4	2,086.9	18,877.1	22,016.4
2.20	354.7	1,868.2	2,198.6	19,369.5	20,684.6
2.40	298.0	1,974.3	2,217.1	18,895.9	18,032.1

to model drug concentrations in sites of therapeutic effect and of toxicity makes it possible to predict the most advantageous ways of delivering drugs.

The final example of the use of PHYSIOPLEX in PB-PK modeling illustrates its ability to model metabolites as well as the parent drug. When a simulation is run of a drug that is susceptible to liver metabolism, the program creates a disk file of amount of metabolite produced as a function of time. Then, in a subsequent run, a parameter file containing information on the metabolite's tissue–plasma partition ratios and its elimination rate constants is read, the program is instructed to use the previously calculated metabolite production rates as its input, and the concentration–time profiles for the metabolite can be then predicted for the various tissues.

An example of a PB-PK simulation of tissue disposition of an Adriamycin metabolite, adriamycinol, is shown in Table 2.3. The highest levels of the metabolite at early time points occur in liver, which is not surprising, since it is being produced in that organ. At later times, gut lumen contained the highest levels, a result of biliary

Table 2.3   PHYSIOPLEX: Physiological Pharmacokinetic Model Modeling
            Adriamycinol in Rabbit

Time	Plasma	Muscle	Adipose	Tumor	Brain
0	0	0	0	0	0
4.00	0.51	1.23	0.20	0	0.09
8.00	0.49	1.43	0.20	0	0.09
12.00	0.43	1.29	0.17	0	0.08
16.00	0.35	1.08	0.14	0	0.06
20.00	0.29	0.88	0.12	0	0.05
24.00	0.23	0.72	0.09	0	0.04

Beta-phase half-life = −2.38
Gamma-phase half-life = 13.5

Time	Kidney	Bone marrow	Liver	GI mucosa	Gut lumen
4.00	3.31	0.38	9.08	0.51	4.53
8.00	3.25	0.37	9.19	0.50	9.00
12.00	2.79	0.32	7.98	0.43	10.68
16.00	2.31	0.26	6.61	0.35	11.14
20.00	1.88	0.22	5.39	0.29	10.33
24.00	1.53	0.18	4.39	0.23	9.12

Time	Lung	Spleen	Heart	Remainder
4.00	0.77	1.01	0.51	0.51
8.00	0.78	0.99	0.49	0.50
12.00	0.67	0.85	0.43	0.43
16.00	0.59	0.70	0.35	0.36
20.00	0.46	0.57	0.29	0.29
24.00	0.37	0.47	0.24	0.24

Drug eliminated in urine = 43.2%
Drug eliminated in feces = 30.0%
Drug metabolized = 0%
Renal clearance = 3.63 ml/hr/gm

secretion of the metabolite. Note that, unlike the parent drug, most of the metabolite was eliminated in urine and feces within 24 h.

Note: Although published too late to be discussed above, a recent article is required reading: S. B. Charnick et al., *J. Pharmacokinet. Biopharm.*, 23: 217–229 (1995) discuss the contribution of PB-PK modeling to drug development from an industry perspective. See also the response to that paper by a group of FDA staff (Ludden et al., *J. Pharmacokinet. Biopharm.*, 23: 231–235, 1995).

The ability to do PB-PK modeling is particularly useful when the metabolite is pharmacologically active (as with the amiodarone example discussed above); in such cases, the pharmacodynamic contributions of both parent and metabolite need to be taken into account.

CHAPTER 3

Interspecies Scaling

I. INTERSPECIES SCALING BY ALLOMETRY

If we ponder the difference between a human and a rat from the perspective of a pharmacologist, the first thing that generally strikes us is that a human is about 300 times larger (by body weight). However, it is not generally the case that a human dose of drug will be 300 times the rat dose. As a first approximation, observation suggests that drug doses tend to scale across species in proportion with body surface area, rather than body weight. A 70-kg human has a surface area of about 1.8 m^2, and a 0.240-kg rat has a surface area of about 0.034 m^2, so our second approximation suggests that the human dose should be about 50 times the rat dose. In other words,

$$D = a \cdot W^{0.67} \tag{3.1}$$

where D is drug dose in a particular individual (of any species), a is a constant, and W is body weight. This is an example of an *allometric equation*, which asserts that many biological parameters (Y) relate across individuals or species by a power law:

$$Y = a \cdot W^b \tag{3.2}$$

(Krishnan and Andersen, 1991; see also Dedrick, 1973; Boxenbaum, 1982; Mordenti, 1986; Yates and Kugler, 1986). Common sense tells us that a human and a rat differ in more than just size, that proportions of body parts also differ; in other words, the allometric constant b will not always equal 0.67. A number of literature values of a and b for various physiological parameters are summarized in Table 3.1. Krishnan and Andersen (1991) distinguish five general cases:

1. Parameters that decrease as body weight W increases; i.e., b < 0. An example is the number of heartbeats per unit time, for which b = –0.25.
2. Parameters that are independent of W, i.e., b = 0. An example is body core temperature.

Table 3.1   Allometric Constants for Various Physiological Processes

Process, Y	Units	a	b	Reference
Water intake	ml/h	0.01	0.88	Adolph, 1949
Urine output	ml/h	0.0064	0.82	Adolph, 1949
O_2 consumption	ml(STP)/h	3.8	0.734	Adolph, 1949
Breath time	s	0.169	0.28	Boxenbaum, 1982
Heartbeat time	s	0.0428	0.28	Boxenbaum, 1982
Creatinine clearance	ml/h	4.2	0.69	Boxenbaum, 1982
Urea clearance	ml/h	1.59	0.72	Adolph, 1949
Inulin clearance		1.74	0.77	Adolph, 1949
Metabolic rate	kcal/d	0.394	0.75	Boxenbaum and Ronfeld, 1983
Life span	year	2.823	0.2	Yates and Kugler, 1986
Body surface area	cm²	12.564	0.64	Yates and Kugler, 1986
Organ weights	g			Adolph, 1949
Kidneys		0.0212	0.85	
Brain		0.081	0.7	
Heart		0.0066	0.98	
Lungs		0.0124	0.99	
Liver		0.082	0.87	
Thyroids		0.00022	0.80	
Adrenals		0.0011	0.92	
Pituitary		0.00013	0.76	
Stomach + intestines		0.112	0.94	
Blood		0.055	0.99	
Plasma		0.0429	0.992	Mordenti, 1986
Muscle		0.463	1.009	
Gut lumen		0.0578	0.934	
Spleen		0.00452	0.901	
Plasma flow rates	ml/min			Mordenti, 1986
Plasma		0.601	0.778	
Muscle		0.0817	0.81	
Kidneys		0.0845	0.802	
Liver		0.0944	0.792	
Gut		0.102	0.787	
Heart		0.00577	0.965	
Lungs		0.215	0.82	
Spleen		0.00179	1.028	
Bone marrow		0.0164	0.804	

Note: The values a and b are coefficients of the equation:

$$Y = a \cdot W^b$$

(body weight W in grams; units of Y as shown above).

3. Parameters that increase with, but not as fast as, W, i.e., $0 < b < 1$. Examples are blood volume circulation time ($b = 0.21$), metabolic rate ($b = 0.75$), and hepatic blood flow ($b = 0.89$).
4. Parameters that increase in proportion with W, i.e., $b = 1$. Examples include many organ weights, e.g., heart, lung, spleen gut, blood, and muscle.
5. Parameters that increase with, but faster than, W, i.e., $b > 1$; e.g., bone weight ($b = 1.083$).

If we have estimates of classical PK parameters in several species (e.g., mouse, rat, dog, and monkey), we can calculate allometric constants for volume of distribution, beta-phase plasma half-life, etc. and use these constants to predict the PK constants for humans. Doing this involves many uncertainties, but is certainly preferable to guesswork.

Dedrick et al.(1970) proposed that many PK parameters were more directly comparable between species if they were compared on the basis of "physiological time" instead of chronological time. Physiological time is defined as

$$t_p = t/W^{0.25} \qquad (3.3)$$

On this time scale, all species have the same heart rate. If we assume that the allometric constant for volume of distribution = 1, and for clearance = 0.75, then (on a scale of physiological time) AUC will be the same for all species.

A. Volume of Distribution

Boxenbaum (1984) related steady state volume of distribution (V_d) to body volume as follows:

$$V_d = V_p + V_T(f_u/f_{uT}) \qquad (3.4)$$

where V_p = plasma volume, V_T = total volume of the extracellular compartments into which the drug distributes, f_u = unbound fraction of drug in plasma, and f_{ut} = weighted mean unbound fraction of drug in tissues. For drugs where the steady-state volume of distribution is much greater than the plasma volume and f_u does not vary between species,

$$V_d = a_1(V_T/f_{uT})^b \qquad (3.5)$$

where a_1 is a constant of proportionality and b is an allometric constant (Krishnan and Andersen, 1991); these authors point out that b tends to cluster around unity, making the volume of distribution proportional to the body weight, i.e.,

$$V_d = a_1 \cdot W^{1.0} \qquad (3.6)$$

Some experimental values for the allometric coefficient for V_d are methotrexate, b = 0.92; cytoxan, 0.99; antipyrine, 0.96; azthreonam, 0.91 (Boxenbaum, 1984); cefoperazone, 0.91 (Ings, 1990), and the HIV protease inhibitor AG1343, b = 0.85 (B. Shetty, personal communication). While these values are close to unity, they are all less than unity, suggesting on the basis of this small sample that V_d may not quite scale exactly in proportion with body weight. Note that if V_d is expressed as liters per kilogram, rather than as liters, the values of b will equal the cited values minus one.

B. Clearance

As a reminder, plasma clearance, Cl_p, is defined as the volume of plasma cleared of drug per unit time; it may also be defined as

$$Cl_p = D/AUC \tag{3.7}$$

Krishnan and Andersen (1991) give the allometric relationship:

$$Cl_p = a_2 \cdot W^{0.75} \tag{3.8}$$

Experimentally determined allometric coefficients for a variety of agents tend to cluster around this theoretical value of 0.75; e.g., for inulin, b = 0.77 (Adolph, 1949) or 0.72 (Boxenbaum, 1984); methotrexate, 0.69 (Dedrick et al., 1970); antipyrine, 0.885 (Boxenbaum, 1980); cytarabine, 0.79 (Dedrick et al., 1973); cytoxan, 0.75 (Boxenbaum, 1984); 2-amino-1,3,4-thiadiazole, 0.73 (King and Dedrick, 1979); phenytoin, 0.915 (Boxenbaum, 1984); 2,3,7,8-tetrachlorodibenzofuran, 0.69 (Boxenbaum, 1984); azthreonam, 0.66 (Ings, 1990); cefoperazone, 0.57 (Ings, 1990); p-aminohippurate, 0.80 (Adolph, 1949), azidothymidine, 0.96 (Patel et al., 1990) and vinyl chloride, 0.75 (Boxenbaum, 1984). For drugs where total clearance is primarily through liver metabolism, it is likely that clearance will be proportional to hepatic blood flow, for which b = 0.80.

For a classical one-compartment model,

$$C_t = C_0 \cdot \exp(-\beta t) \tag{3.9}$$

where C_t is plasma concentration at time t, C_0 is initial concentration after an i.v. bolus dose = D/V_d, and β is the first order elimination rate constant. This rate constant is related to clearance by

$$\beta = Cl_p/V_d \tag{3.10}$$

which means that

$$\beta = a_2/a_1 \cdot W^{0.75}/W^{1.0} \tag{3.11}$$

Since elimination-phase half-life is related to elimination rate constant by

$$t_{1/2} = 0.693 \cdot V_d/Cl_p \tag{3.12}$$

it follows that

$$t_{1/2} = 0.693 \cdot a_1/a_2 \cdot W^{1.0}/W^{0.75} \tag{3.13}$$

so that

$$t_{1/2} = 0.693 \cdot a_1/a_2 \cdot W^{0.25} \tag{3.14}$$

which means that (other things being equal) we expect plasma half-life in a 65-kg human to be about four times longer than in a 250-g rat (and half-life in a 250-g rat to be 1.8 times longer than in a 25-g mouse). To express this conclusion in the "physiological time" frame, $t_{1/2}$, like AUC, corresponds to an equal number of heartbeats in different species. Boxenbaum (1982) discussed actual data for eight drugs where the mean allometric coefficient for elimination-phase half-life was 0.222 (methotrexate, 0.228; cytoxan, 0.236; antipyrine, 0.069; digoxin, 0.234; hexobarbital, 0.348; phenylbutazone, 0.060; aniline, 0.176; and diazepam, 0.428). Other values are ceftizoxime, 0.248 (Mordenti, 1986); azthreonam, 0.186; and cefoperazone, 0.212 (Ings, 1990). Ings (1990) quotes data for nine β-lactam antibiotics for which the mean b value is 0.249.

Renal clearance, i.e., drug elimination in urine, will only equal plasma clearance for drugs that are eliminated entirely in the urine. For such drugs, renal clearance should follow an allometric relationship similar to Equation 3.4, with clearance proportional to the three-fourths power of body weight. According to Krishnan and Andersen (1991), experimental values of the allometric exponent for a number of drugs were between 0.69 and 0.89, in rough agreement with the expected value of 0.75. Dedrick et al. (1970) point out that the clearance of methotrexate is a constant fraction of the creatinine clearance in a range of species, and this is apparently the case for a number of other drugs that are eliminated primarily by the kidneys.

Hepatic clearance (which may be caused by elimination of unchanged drug into bile, or by metabolic conversion) appears to vary much more erratically between species than renal clearance. If a drug is eliminated unchanged into the bile and this rate is blood flow limited, then hepatic clearance should follow an allometric equation with exponent 0.8. For compounds eliminated primarily by metabolism, the rate of hepatic clearance will depend upon the K_m and V_{max} values for the metabolic enzymes, and these do not alter in any systematic way between species. They may, however, be measured *in vitro*, and this is discussed in Chapter 4.

C. Measuring Allometric Constants Experimentally

If we want to predict human pharmacokinetic parameters from preclinical data, a more accurate procedure than simply using theoretical allometric coefficients is to obtain PK data in several animal species and then calculate the allometric constants. Mordenti (1986) suggests the following procedure:

1. Determine PK parameters experimentally, preferably in four species (e.g., mouse, rat, rabbit, and dog).
2. Perform linear regression on the relationship log Y (where Y is the PK parameter) vs. log W (W is body weight in grams). This will give the equation:

$$\log Y = a' + b \cdot \log W \tag{3.15}$$

and taking antilogs gives $Y = 10^{a'} \cdot W^b$, which, if we define $a = 10^{a'}$, is equal to Equation 3.2.

An experimental example of fitting data to the allometric equation is shown in Figure 3.1. This plots $\log(V_d)$ against $\log(W)$ for four species, rat, marmoset, cynomolgus monkey, and dog. Fitting the best straight line to the log–log plot by linear regression gives the relationship:

$$\log(V_d) = 0.546 - 0.1478 \log(W) \tag{3.16}$$

or, alternatively,

$$V_d = 3.516 \cdot W^{-0.1478} \tag{3.17}$$

For a 70-kg human, this predicts a volume of distribution of 1.88 l/kg. The value of V_d estimated from the phase I clinical trial was 1.5 l/kg (B. Shetty, personal communication).

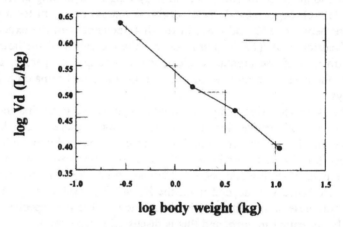

Figure 3.1 Volumes of distribution of the antiviral agent AG1343 in four animal species. The slope of the straight line fitted to these experimental data gives an allometric constant that may then be used to predict values of V_d in other species.

D. Neoteny

Allometric rules provide a reasonable first guess of what sort of PK parameters to expect in humans, based upon preclinical data, but there are many exceptions. Many of the exceptions, as noted above, are the result of interspecies differences in drug metabolism. However, it is clear that humans are outliers in a number of respects; e.g., using Table 3.1 to predict brain size of a 70-kg human gives a value of 200 g, low by about a factor of seven. This and other anomalies of human anatomy are often attributed to neoteny, the retention of fetal characteristics in the adult. Yates and Kugler (1986), as evidence for retarded development in humans, cite the fact that humans reach puberty at about 60% of their final body weight, while most laboratory and farm animals do so earlier, at only 30% of their final weight. Other evidence for neoteny comes from morphological comparisons of, for example, adult human skulls, which resemble in general proportions the skull of a fetal ape much more than that of an adult ape (Boxenbaum and D'Souza, 1987). Humans are outliers in other respects than anatomical proportions, a notable example being maximum life span potential (MLP). Based upon the allometric parameters in Table 3.1, humans should have an MLP of about 26 years. Whether longer life span is connected in any way with larger brain size is not entirely clear, though it has been suggested that greater neural complexity, delay of maturity, and life span are interconnected phenomena.

Sacher (1959) has shown that an allometric equation that includes an expression for brain size as well as for body weight gives a reasonable estimate of MLP across a wide range of species, including humans:

$$MLP = 10.839 \cdot B^{0.636} \cdot W^{-0.225} \tag{3.18}$$

where B is brain weight (grams). This equation suggests a human MLP of about 90 years. Campbell and Ings (1988) have shown that if brain size is factored in, allometric equations can give reasonable estimates of clearance for drugs that are eliminated primarily through the hepatic mixed-function oxidase system, e.g., antipyrine and phenytoin:

$$CL_p = a \cdot B^c \cdot W^b \tag{3.19}$$

Boxenbaum (1982) showed that when clearance of antipyrine and phenytoin was expressed as liters per MLP, rather than liters per hour, then a plot of clearance vs. body weight was linear on a log–log scale (Figure 3.2). The same article gives values of MLP for 11 mammalian species. This result suggests a general relationship between life span (or to put it another way, physiological time) and metabolic rate.

E. How Useful Is Allometry in Pharmacokinetics?

Allometry is best regarded as a rough-and-ready way of arriving at some reasonable first approximations, based upon the preclinical data, of what human PK values may reasonably be expected to be. The simplest and roughest approximation comes from using Equations 3.6, 3.8, and 3.14, which can give estimates of V_d, CL_p, and $t_{1/2}$, respectively. Using the case of predicting volume of distribution of the HIV protease inhibitor AG1343 as an example, if we had applied Equation 3.6 to the rat data, the predicted volume of distribution for a 70-kg human would be 300.3 l (compared to the experimental vaue of 105 l), an error of 186%; applying Equation 3.6 to the dog data gives V_d for a 70-kg human of 172.9 l, an error of 65%. If preclinical data are available for multiple species, which was the case for AG1343, fitting an allometric equation to preclinical data from multiple species, as described above, may give a more accurate prediction. For the example of AG1343, the allometric constants obtained from preclinical data in four animal species gave an estimated V_d for a 70-kg human of 131.6 l (25% high). While there is no way of telling how typical this single example may be, it may give some idea of what to expect. Of course, obtaining preclinical data with four species is not a trivial undertaking, but, as with AG1343, it is not particularly unusual to have such data before

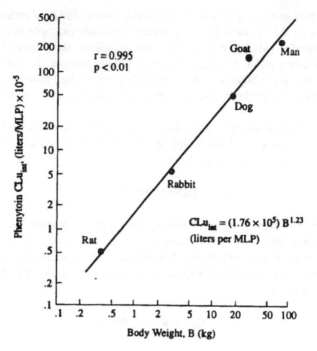

Figure 3.2 Allometric relationship between unbound phenytoin intrinsic clearance per maximum life span potential and body weight. (From Boxenbaum, 1982. Reproduced with permission of Plenum Press.)

making a decision to take a drug into human trials. Bachmann (1989) gives other examples of prediction of human V_d, clearance, and half-life from data in rat, dog, and monkey.

II. INTERSPECIES SCALING BY PB-PK MODELING

Allometric scaling is simple, but empirical. It is most usefully employed, as described above, to predict classical PK parameters, such as elimination-phase half-life or volume of distribution across species. If we wish to make interspecies extrapolations that involve drug concentrations or drug kinetics in tissues other than plasma, or if we want to ask mechanistic questions concerning drug distribution, or pharmacodynamic questions about a drug that acts in a particular organ, we must use a PB-PK model. Similarly, if we have some *in vitro* information that we want to incorporate into our interspecies scaling, we can use a PB-PK model. Comparative measurements of *in vitro* metabolism of a test compound by mouse and human microsomes cannot be incorporated into an allometric estimate of the compound's plasma half-life, but it can be easily incorporated into a PB-PK model.

Since developing a PB-PK model for drug disposition in humans directly is very difficult, scaling from test data in laboratory animals is often the only way to construct such a model. Sometimes the model can be partially validated with *in vitro* values for tissue–plasma partition coefficients obtained from autopsy samples. Even more rarely, biopsy samples may be obtainable for one or more human tissues.

Tables 3.2 and 3.3 summarize organ weights and blood flows for eight species. Many of these values are based on a paper by Davies and Morris (1993), others are from Gerlowski and Jain (1983), and any gaps have been filled in by allometric scaling. Using these values, if we have a physiological pharmacokinetic model of distribution of an experimental drug in one species (e.g., rat), we can substitute the tissue volumes and flow rates for another species (e.g., human) into our model and come up with an estimate of the PK properties of our drug in the second species. For background reading on this topic, readers are referred to a classic review by Rowland (1985), and more recent articles by Krishnan and Andersen (1991), Kawai et al.(1994), and Björkman et al. (1994). Obviously, interspecies scaling by PB-PK modeling is more computer intensive than allometric scaling. However, by using a program package such as PHYSIOPLEX, it is not necessary to do any reprogramming. This means that the additional investment in preclinical drug development in principle can be limited to the work required to produce a PB-PK model of the developmental drug in a single species (usually rat). It is argued in the previous chapter that a rat PB-PK model should be a standard part of a drug's preclinical workup in any case.

Table 3.2 Organ Volumes in Eight Mammalian Species

Parameter	Mouse	Hamster	Rat	Guinea pig	Rabbit	Monkey	Dog	Human
Body wt(g)	22	150	250	400	2,500	5,000	10,000	70,000
Cardiac output (ml/h)	488	2,440	3,490	5,113	22,600	39,550	69,300	336,000
Tissue volumes (ml)								
Plasma	1.0	6.5	11.1	17.7	108	215	426	2,920
Muscle	9.7	68.7	111.2	178	1,350	2,500	5,530	35,000
Fat	0.52	5.4	10.2	18.2	174	407	955	10,460
Brain	0.39	1.4	1.8	2.7	14	90	80	1,400
Kidney	0.34	1.4	2.0	3.4	13	25	50	310
Bone marrow	0.6	3.5	5.6	9.3	47	135	120	1,400
Liver	1.3	6.9	10.6	16.2	77	150	320	1,800
Intestines	1.5	12.2	11.3	17.4	120	230	480	2,100
Gut lumen	1.5	7.0	8.8	15.6	86	230	314	2,100
Lungs	0.12	0.74	1.5	2.6	18	33	100	1,000
Spleen	0.1	0.54	1.3	1.4	1	8	25	180
Heart	0.1	0.6	1.1	1.9	5	18.5	80	330

Early examples of interspecies scaling by physiological modeling include the scaling from dog to human of thiopental kinetics (Bischoff and Dedrick, 1968), of methotrexate from rat to human (Bischoff et al., 1970) and of lignocaine kinetics from monkey to human (Benowitz et al., 1974). Rowland (1985) commented that the success seen with the methotrexate scaling may be attributable to the fact that it is primarily cleared by glomerular filtration, and the success with lignocaine may stem from its rapid hepatic clearance, which is perfusion-rate limited. Both perfusion and glomerular filtration are essentially independent of the drug under study. Rowland

Table 3.3   Blood Flow Rates in Eight Mammalian Species

Tissue	Mouse	Hamster	Rat	Guinea pig	Rabbit	Monkey	Dog	Human
Muscle	54.6	269	450	552	2,410	5,400	7,320	34,900
Fat	10.3	46	24	125	1,920	1,200	2,100	11,000
Brain	31.8	87	78	191	579	2,770	2,700	42,000
Kidney	78	372	552	648	4,800	8,280	12,960	53,100
Bone marrow	10.2	69	136	199	660	1,380	1,200	7,200
Liver[a]	108	390	828	960	4,442	7,930	18,540	87,000
Intestines	90	318	450	599	2,705	4,790	12,960	66,000
Lungs[b]	488	2,440	3,490	5,113	22,600	39,550	69,300	336,000
Spleen	5.4	15	18	83	540	1,260	1,500	4,620
Heart	16.8	8.4	234	284	960	1,840	3,240	14,400
Urine flow	0.05		2.1		6.3	15.6	12.5	58.3
Bile flow	0.08		0.9		12.5	5.2	5	14.6

Note: (1) All rates are expressed as ml/h. (2) Plasma flows may be calculated as 58% of whole
    blood flow.
[a] Including portal circulation.
[b] Total cardiac output.

predicted that species scaling by PB-PK modeling would be more difficult for drugs
whose extraction ratio was low and dependent on enzymatic or active transport-
mediated processes, which may show marked differences between species.

   To illustrate the process of interspecies scaling by PB-PK modeling, we will
consider the example of Adriamycin (doxorubicin) in the rabbit, discussed at length
in Chapter 2. Output from the model of Harris and Gross (1975), as emulated by
the PHYSIOPLEX program, is shown in Table 2.1. Let us now instruct PHYS-
IOPLEX to load the human parameter set, and rerun the simulation. This output is
shown in Table 3.4. It differs from the rabbit data in a number of respects: tissue
levels for liver at any particular time are higher in the (simulated) human than for
rabbit, and for muscle the levels are lower at all times in human. For some other
tissues (brain, kidney, and gut) levels start out higher in the human, but end up lower,
while, conversely, the predicted level for human bone marrow starts lower than for
rabbit, but ends up higher. Also, the terminal half-life is longer in human. It appears
that simply substituting human organ volumes and blood flow rates is not enough
to confer human behavior on the model: further adjustments are necessary. First, we
should remember that doses of anticancer drugs are customarily scaled in proportion
to body surface area, not body mass. If we assume that the surface area of a 2.5-kg
rabbit is 0.1875 m^2, then a 3-mg/kg dose equals 40 mg/m^2. For a 1.8-m^2, 70-kg
human, this is 72 mg, or 1.03 mg/kg. Table 3.5 shows a simulation run at this dose,
which now gives plasma levels at early time points that are unrealistically low. In
general, both rate constants for renal excretion and for hepatic metabolism are lower
for human than for smaller species. As a first approximation we can assume that the
renal excretion rate constant is proportional to the ratio of creatinine clearance to
kidney blood flow. The prediction of hepatic metabolism in different species is
discussed in Chapter 4. For now, we will simply assume that the rate constant for
metabolism of Adriamycin in human liver is 40% of the rate in rabbit liver. Substituting

Table 3.4 PHYSIOPLEX: Physiological Pharmacokinetic
Model Modeling Adriamycin in Human

		3-mg/kg i.v. Bolus			
Time	Plasma	Muscle	Adipose	Tumor	Brain
0.0	132,201.9	0.0	0.0	0.0	0.0
4.0	135.8	1,752.6	1,384.0	0.0	1,866.4
8.0	69.6	1,855.5	1,089.8	0.0	845.1
12.0	45.8	1,797.8	798.9	0.0	458.2
16.0	34.0	1,687.0	580.8	0.0	300.3
20.0	27.0	1,558.5	427.7	0.0	222.7
24.0	22.2	1,426.9	322.0	0.0	176.8

Alpha-phase half-life = 0.111
Beta-phase half-life = 2.73
Gamma-phase half-life = 15.1

Time	Kidney	Bone marrow	Liver	GI mucosa	Gut lumen
4.0	102,976.3	6,643.2	9,030.6	8,193.9	902.1
8.0	45,580.6	5,137.6	3,657.4	3,194.1	1,143.9
12.0	24,721.1	3,708.4	2,229.5	1,920.2	1,246.3
16.0	16,319.4	2,663.2	1,605.4	1,375.2	1,301.2
20.0	12,172.5	1,944.4	1,251.8	1,069.6	1,332.5
24.0	9,696.5	1,456.3	1,019.2	869.7	1,349.3

Time	Lung	Spleen	Heart	Remainder
4.0	15,020.6	43,017.4	7,149.7	1,829.6
8.0	8,058.8	42,045.6	3,177.2	1,405.3
12.0	5,420.0	38,131.9	2,020.8	1,020.4
16.0	4,072.9	33,650.7	1,480.6	738.7
20.0	3,245.1	29,325.2	1,164.2	543.3
24.0	2,675.4	25,395.1	952.4	409.3

Drug eliminated in urine = 2.2%
Drug eliminated in feces = 0%
Drug metabolized = 77.8%
Renal clearance = 1.03 ml/h/g
Plasma AUC = 2.88 μM/h

these values into the model gives us the results shown in Table 3.6, which, in the absence of any clinical data, represents our best estimate of how Adriamycin is disposed in humans.

Figure 3.3 plots the predicted cardiac levels of Adriamycin for the three PB-PK models. The heart level of Adriamycin is presumed to predict for the cumulative cardiomyopathy caused by this drug. Based upon these predictions, a dose of 1.03 mg/kg Adriamycin in humans should be less cardiotoxic than the 3-mg/kg dose in rabbits.

Once we have some clinical data, we have the opportunity to "tweak" the parameters of the model to bring it closer to reality. Although we cannot usually get data on human tissue levels, we do usually have plasma levels, and patterns of elimination of drug and metabolites in urine and feces are measurable. The rates of urinary and biliary excretion and of hepatic metabolism can thus be fitted to the experimental data.

Table 3.5 PHYSIOPLEX: Physiological Pharmacokinetic Model Modeling Adriamycin in Human

1.03-mg/kg I.V. Bolus

Time	Plasma	Muscle	Adipose	Tumor	Brain
0.0	45,389.3	0.0	0.0	0.0	0.0
4.0	46.6	601.7	475.2	0.0	640.7
8.0	23.9	637.0	374.1	0.0	290.0
12.0	15.7	617.3	274.3	0.0	157.4
16.0	11.7	579.3	199.5	0.0	103.1
20.0	9.3	535.2	146.9	0.0	76.5
24.0	7.6	490.0	110.6	0.0	60.7

Alpha-phase half-life = 0.111
Beta-phase half-life = 2.73
Gamma-phase half-life = 15.1

Time	Kidney	Bone marrow	Liver	GI mucosa	Gut lumen
4.0	35,348.0	2,280.7	3,099.6	2,812.5	309.7
8.0	15,643.7	1,763.6	1,255.3	1,096.3	392.8
12.0	8,489.3	1,273.4	765.6	659.4	427.9
16.0	5,605.7	914.7	551.4	472.4	446.7
20.0	4,181.1	667.9	430.0	367.4	457.5
24.0	3,330.3	500.2	350.0	298.7	463.3

Time	Lung	Spleen	Heart	Remainder
4.0	5,750.5	14,621.0	2,894.2	641.6
8.0	2,947.3	14,559.1	1,176.4	501.2
12.0	1,941.7	13,278.2	727.5	365.0
16.0	1,444.0	11,745.8	526.2	264.0
20.0	1,143.9	10,247.9	410.9	193.7
24.0	939.8	8,879.6	334.8	145.4

Drug eliminated in urine = 2.2%
Drug eliminated in feces = 0%
Drug metabolized = 77.8%
Renal clearance = 1.03 ml/h/g
Plasma AUC = 2.88 µM/h

A survey of the PB-PK modeling literature gives numerous examples of the use of such models for interspecies scaling. An example mentioned in Chapter 2 was the use by Chan et al. (1978) of the Harris and Gross model to predict Adriamycin tissue distribution in humans. An even earlier model, that of Dedrick et al. (1973), which predicted cytarabine distribution in four species, was of particular interest in that it incorporated *in vitro* data on drug metabolism for the various species. Some particularly interesting recent examples of the use of PB-PK models in interspecies scaling are the model of Farris et al. (1988) for cisplatin, the model of the nitrosourea anticancer drug ACNU (Mitsuhashi et al., 1990), the model of Björkman et al. (1994) of fentanyl and alfentanil, and the model of Kawai et al. (1994) describing kinetics of the cyclosporin analog SDZ IMM 125 in rats, dogs and humans. These and the other models discussed in Chapter 2 make it clear that, despite the difficulties of the technique, PK-PK modeling can often provide useful insights into the bodily disposition of drugs of many classes in humans.

Table 3.6   PHYSIOPLEX: Physiological Pharmacokinetic Model Modeling
            Adriamycin in Human

1.03 mg/kg I.V. Bolus

Time	Plasma	Muscle	Adipose	Tumor	Brain
0.0	45,389.3	0.0	0.0	0.0	0.0
4.0	70.4	666.4	534.5	0.0	758.8
8.0	40.1	765.2	470.5	0.0	418.8
12.0	27.7	778.3	373.3	0.0	257.1
16.0	21.1	756.4	289.7	0.0	180.1
20.0	17.1	718.9	225.9	0.0	138.6
24.0	14.4	674.5	179.0	0.0	113.0

Alpha-phase half-life = .114
Beta-phase half-life = 5.40
Gamma-phase half-life = 17.1

Time	Kidney	Bone marrow	Liver	GI mucosa	Gut lumen
4.0	42,958.6	2,568.5	7,577.5	3,629.5	522.2
8.0	23,493.5	2,225.1	3,605.5	1,766.3	752.4
12.0	14,408.8	1,742.2	2,300.3	1,142.5	866.8
16.0	10,116.9	1,338.1	1,694.4	846.9	936.0
20.0	7,803.1	1,035.9	1,345.9	675.1	981.6
24.0	6,367.5	817.3	1,116.3	561.0	1,012.6

Time	Lung	Spleen	Heart	Remainder
4.0	8,688.1	16,240.4	3,861.5	722.1
8.0	4,952.2	17,637.3	1,923.8	631.2
12.0	3,414.9	16,977.9	1,265.6	497.9
16.0	2,611.6	15,640.5	945.2	384.6
20.0	2,116.9	14,117.8	756.1	298.9
24.0	1,776.6	12,610.1	629.6	236.4

Drug eliminated in urine = 1.7%
Drug eliminated in feces = 0.1%
Drug metabolized = 67.5%
Renal clearance = .625 ml/h/g
Plasma AUC = 1.3 μM/h

III. SPECIES DIFFERENCES IN EXTRACELLULAR AND INTRACELLULAR BIOCHEMISTRY

The cytarabine model (Dedrick et al., 1973) showed the value of the PB-PK technique in incorporating *in vitro* biochemical data for appropriate species into a model. In this particular case the enzymes concerned catalyzed intracellular reactions of drug activation and of turnover of active drug species. The kinetic parameters (K_m and V_{max}) for such reactions often vary, apparently quite erratically, between species. Cytidine deaminase (which deaminates and inactivates cytarabine), e.g., is very active in mouse tissues, but less so in the rat. In humans, activity is closer to that of the mouse than the rat, so the mouse is a more appropriate test species than the rat for drugs that are deaminated by this enzyme.

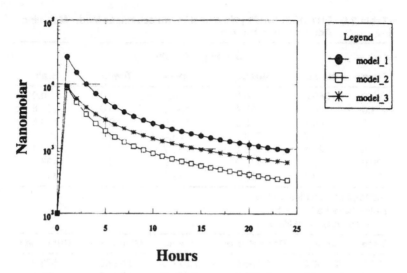

Figure 3.3 Cardiac Adriamycin levels predicted by PB-PK modeling. Model 1 is the data of Table 3.4, model 2 is the data of Table 3.5, and model 3 shows the data of Table 3.6.

Of course, it is not only enzymes in the target tissue that influence drug disposition. Enzymes of drug metabolism in the liver also vary widely in total activity (indeed in presence or absence) and in K_m between species. Predicting and modeling hepatic drug metabolism are discussed in Chapter 4. Plasma biochemistry, too, shows interspecies differences that can exert a major effect on drug efficacy and toxicity. For instance, inhibitors of the *de novo* pyrimidine biosynthetic pathway are antagonized by circulating uridine. Mice have much lower circulating uridine concentrations than humans, so that *in vivo* experiments in mice tend to overpredict for the activity of such drugs (e.g., brequinar) in humans. Conversely, the thymidylate synthase inhibitor class of anticancer agents (e.g., Tomudex) is antagonized by plasma thymidine, which has much higher concentrations in mice than in humans, so that for this class of drugs, murine test systems underpredict for human activity. This kind of idiosyncratic biochemical difference between species clearly influences our choice of preclinical test species and the interpretation of results. In principle, we can factor in such biochemical differences into pharmacodynamic models, if those models are sufficiently sophisticated. Pharmacodynamic models that include detailed biochemical descriptions of drug action and software for this kind of modeling are discussed in Chapter 5.

Another part of the drug development process where we are forced to draw conclusions about likely human responses based upon data in other species is, of course, toxicology. Which are the "right" species for preclinical pharmacology and toxicology? This question is discussed by Fitzgerald (1993), and her answer is that the right species is the one that most closely resembles humans in distribution and metabolism of the particular drug. She discusses various examples of toxicities that occur in one species, but not in another, and shows that such toxicities can often be related to species-specific accumulation of drug at particular sites. In such cases, the advantage of having a model of drug distribution in humans is very clear.

Leal et al. (1993) discuss the use of allometry in toxicokinetics, i.e., the relationship of drug concentration (particularly plasma concentration, in this example) to toxicity and its variation between species. Rather than relying too much upon a single species, they argue for basing clinical study design upon allometric scaling of data from a number of species. In extrapolating animal data to humans, they argue for the concept of the safe/toxic plasma concentration range, as distinct from the safe/toxic dose range, and a knowledge of whether it is the peak plasma concentration, the steady state plasma concentration, or the AUC, that best correlates with toxic or therapeutic effects.

Species differences are also discussed by Yacobi et al. (1993). They make a number of recommendations: (1) PK and ADME studies should be done in at least three species (selected from mouse, rat, rabbit, dog, and monkey) to allow for allometric scaling. (2) Studies allowing for interspecies comparison of metabolism profiles should be performed, including *in vitro* studies that would allow for qualitative comparison of humans with the animal species used in the toxicology studies. (3) Early *in vivo* metabolism studies should be performed in humans, which would allow comparison of metabolite blood levels between man and toxicology species. (4) Species differences in protein binding should be determined. Any extrapolations of effective or toxic blood levels should include corrections for protein binding.

A particularly informative example of the importance of assessing toxicity in an appropriate species was discussed by Schuplein et al.(1990), who considered, not a drug, but the artificial sweetener, cyclamate. This compound has low oral bioavailability, and the small fraction that is absorbed is quickly eliminated unchanged. However, a small and highly variable fraction of ingested cyclamate is converted by the intestinal flora into cyclohexylamine, which is rapidly absorbed. Cyclamate causes testicular atrophy in rats, and this toxic effect is believed to be mediated by cyclohexylamine. Cyclamate does not appear to cause testicular atrophy in mice, and the levels of cyclohexylamine measured in plasma or in testicles following cyclamate feeding are much lower in mice than in rats. Which of these two rodent species is likely to be the most appropriate predictor of human toxicity? Schuplein et al. (1990) argue that the way to answer this question is to measure plasma cyclohexylamine levels in humans who have ingested cyclamate and to compare the results with levels in rats and mice. This again endorses the conclusion that the most appropriate species for toxicology studies is the species whose pharmacokinetics and metabolism of the test substance most closely resemble those of humans. This argument (which is undeniably convincing in the case of cyclamate) rests on the fact that cyclamate concentration in the testis tracks that in the plasma. Plasma concentrations of cyclamate in humans are readily measurable following ingestion, but testicular concentrations are not. If testicular cyclamate levels did not track plasma levels for any reason, a PB-PK model would be essential to relate predicted testicular levels in humans to measured human plasma levels and to the measured testicular levels in rats that correlated with toxicity.

The most compelling case for interspecies scaling rests largely upon the necessity of conducting toxicology in nonhuman species, typically one rodent and one non-rodent species. An attempt is then made to relate particular toxicities to the measured drug concentration in plasma and in the organ or tissue where toxicity is found. If the toxicity is clearly related to the concentration of drug in plasma, then this can easily be measured in humans, making it relatively straightforward to predict the toxicity. If the toxicity is related to the local drug concentration at the site of toxicity, and if this does not have a simple relationship to plasma concentration, then, obviously, predicting that particular toxicity in humans will be less straightforward. It is in this situation that a PB-PK model, used for interspecies scaling, may make a contribution.

IV. PLASMA PROTEIN BINDING IN DIFFERENT SPECIES

Many drugs bind to proteins, both in plasma and in extracellular and intracellular fluid. In plasma, the most common drug-binding protein is plasma albumin, though (particularly for basic drugs) binding to α_1-acid glycoprotein (AAG) is also important. Considering all medically useful drugs, the fraction that is protein bound covers the gamut from essentially zero to greater than 99%.

The topic of binding of drugs to plasma proteins has been reviewed in detail by Kremer et al. (1988). Protein binding is almost always reversible, and saturable, so follows a Michaelis–Menten equation. To estimate the fraction of a drug bound to protein, a useful equation is given on p. 86 of Welling (1986):

$$F\beta = 1/(1 + K/nP + [c]/nP) \tag{3.20}$$

where [c] is free drug, P is protein concentration, n is number of drug-binding sites per molecule of protein, and K is the dissociation constant. The concentration of serum albumin in human plasma is esimated to average 47 mg/ml (= 0.69 mM) and AAG = ~0.9 mg/ml (Wagner, 1993).

The customary approach to PK modeling is to assume that only free (non-protein bound) drug is available for binding to receptor or target enzyme, and similarly, that only free drug is accessible to metabolic enzymes.

If the degree of drug binding to different plasma proteins (albumin, AAG, α_2-globulin, β-globulin, and γ-globulin) is known, then it is possible, by allowing for the different content of these plasma protein fractions in different species, to predict the total plasma binding in a particular species. Mordenti (1986) cites the example of camptothecin, for which the following equation was derived to describe plasma binding across 24 species:

$$\log f_u = 2.12 + 0.0628 \log(\alpha_1) + 0.895 \log(\alpha_2)$$

$$-3.30 \log(\text{albumin}) - 0.651 \log(\text{albumin}) \log(\alpha_1)$$

$$-1.93 \log(\text{albumin}) \log(\alpha_2) \tag{3.21}$$

where f_u is the fraction unbound in plasma (expressed as percent), and albumin, α_1, and α_2 refer to concentrations in plasma (gram per 100 ml) of albumin, α_1-globulin, and α_2-globulin, respectively. For this particular drug, the β-globulin and γ-globulin fractions did not contribute significantly to total protein binding.

The approach of individually characterizing drug binding to the separate plasma fractions, though it is the most rigorous approach, is quite labor intensive. A more common approach is to measure plasma protein binding experimentally (e.g., by equilibrium dialysis), using plasma from the species of interest. Results are often expressed as percent of drug that is protein-bound in plasma. While this approach tells us nothing about which plasma component the drug is actually bound to, it provides sufficient information to model free and bound drug compartments separately in a PB-PK model. In fact, the percent of drug that is protein-bound is not strictly constant, but follows a mass action law. To predict protein binding accurately at any drug concentration requires measurement of the dissociation constant, and this in turn necessitates knowledge of which plasma component the drug is bound to and the concentration of that component in the species of interest.

In a discussion of the pharmacokinetic implications of protein binding of drugs, Welling (1986) points out that for most drugs, the fraction that is bound to protein will be essentially constant over a concentration range of several logs, generally spanning the therapeutic range. For modeling purposes, assumption of a constant fraction of drug bound to plasma protein is often a reasonable approximation.

A. Protein Binding: Boon or Bane?

The fact that a drug is strongly protein-bound is generally considered by drug developers to be an unwelcome nuisance. Since the pharmacodynamically active species is the free drug, the most obvious implication of strong protein binding is that the total plasma concentration of drug needs to be proportionately higher; for a drug that has limited oral bioavailability, this may not be trivial to achieve. Other disadvantages of strong protein binding include the fact that other drugs that may compete for protein-binding sites may cause unexpected, or even dangerous, drug–drug interactions.

There appear to be many popular misconceptions about this topic; Welling (1986) discusses the argument that if a drug binds to plasma protein, then it cannot penetrate tissues. He refutes this argument with the example of erythromycin, which is 90% bound to plasma proteins, but where only 1 to 3% of this drug is found within the plasma, the rest being in the tissues. Another (opposite) misconception is that since many drugs have affinity for their receptor that is in the nanomolar range, while affinity of drugs for plasma proteins is often of the order of 10 μM, that the receptor can simply "strip off" drug molecules from plasma protein. A simple calculation shows the fallacy of this argument. Consider a drug that has 1 nM K_d for its receptor, and is not protein-bound: if we assume that the receptor exists at an effective concentration of 10 nM, the EC_{50} concentration will be 6 nM (based upon the Morrison equation, Equation 5.12). Now assume that the same drug binds to plasma albumin with K_d of 10 μM (four logs weaker than binding to receptor). To achieve an EC_{50} concentration will require a total concentration of 420 nM drug in the plasma, of which all but about 1.5% is protein-bound. Despite the four-log difference in K_d values, the drug still partitions mostly to plasma protein. Why? Because the concentration of plasma albumin is nearly five logs greater than the receptor concentration.

Protein binding may cause large differences in the apparent values of various pharmacokinetic parameters. Welling (1986) shows that the apparent volume of distribution becomes less when a drug is bound to plasma proteins, while it becomes greater when a drug binds to tissue protein.

Does protein binding influence the rate of drug clearance and the AUC? This question is most easily illustrated by a simple modeling study. The model to be used is shown in Figure 3.4. Drug is assumed to exist in two compartments, a central free drug compartment, and a protein-bound compartment. Free drug may interact with its receptor, which for the purpose of this discussion is assumed to exist in the central compartment, and drug is also eliminated directly from the central compartment, at a rate v_2. The drug effect at any instant, v_1, follows a Michaelis–Menten dependence upon free drug concentration, and the total drug effect (AUC) is the integral of v_1 over time. Note that AUC in this instance is pharmacodynamic, i.e., it is the the total drug effect, rather than a concentration–time integral. Let us consider three scenarios for drug elimination: (1) Elimination is first order with respect to free drug concentration. (2) Elimination is primarily metabolic, and follows a saturable (Michaelis–Menten) relationship to free drug concentration, with K_m considerably higher than the K_d for the drug receptor interaction. This is a reasonable assumption, since many drug–receptor interactions have K_d values in the nanomolar range, while reactions of drug metabolism often have K_m values in the micromolar range. (3) This makes the less probable assumption that clearance is metabolic, with $K_m < K_d$.

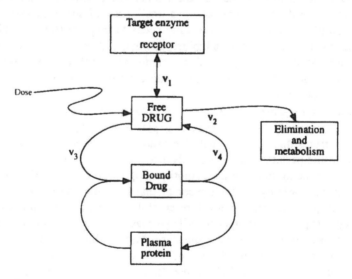

Figure 3.4 Compartmental model used to predict the influence of protein binding on PK parameters.

Case A — Let total drug concentration = 1 μM, K_d for the drug response = 100 nM, and drug elimination be first order, with rate constant = 0.0815 min^{-1}. This predicts a plasma half-life of 8.5 min, assuming no protein binding. As the degree of protein binding is allowed to increase, the half-life becomes longer (Figure 3.5); with 75% protein binding, $t_{1/2}$ = 31.5 min, and with 98% protein binding, $t_{1/2}$ has increased to 430 min. As the degree of protein binding increases, despite the fact that the instantaneous drug effect decreases, the total effect (AUC) increases almost fourfold (Figure 3.6).

Case B — Parameter values are the same as for case A, except that v2 is determined by a Michaelis equation, with K_m = 0.5 μM and V_{max} = 0.1 nmol/ml/min. Again, the $t_{1/2}$ in the absence of protein binding is 8.5 min, and again both the predicted $t_{1/2}$ and the AUC increased as the degree of protein binding increased.

Case C — Parameter values were the same as for case B, except that V_{max} for metabolism was 0.0605 nmol/ml/min and K_m = 0.02 μM. Once again, the $t_{1/2}$ increased as degree of protein binding increased (though much less markedly than for cases A and B), and in this case the AUC decreased as protein binding increased (Figure 3.6).

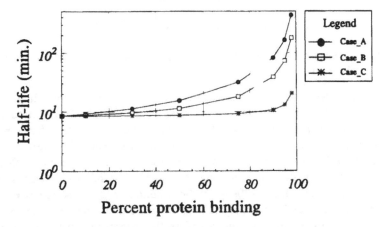

Figure 3.5 Calculated effect of protein binding on plasma half-life. Case A, first order elimination; case B, saturable elimination with K_m for elimination higher than K_d for drug–receptor interaction; case C, saturable elimination with K_m for elimination <K_d.

Protein binding thus can have a major influence on PK parameters. Whether the effect is detrimental or beneficial is obviously a matter of context. However, for those drugs whose effect depends more upon the product of concentration and time than upon a high peak plasma level, the above examples show that in some cases at least protein binding may cause a paradoxical increase in potency.

V. CONCLUSIONS

Allometry is not a computationally intensive technique, requiring only the means to do linear regression analysis. However, it has been reviewed in detail here, because

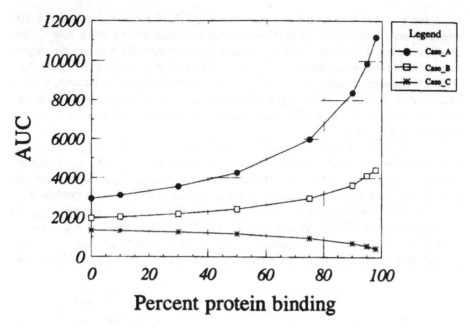

Figure 3.6 Calculated effect of protein binding on AUC. Cases A, B, and C are as defined in the caption of Figure 3.5.

a discussion of the advantages of interspecies scaling by PB-PK modeling requires some grasp of what can be done by allometry. In contrast, PB-PK modeling is unquestionably an approach that could not be undertaken without modern computer power. Indeed, early reviews on the PB-PK modeling approach to interspecies scaling emphasized the large amounts of data needed for these models, the complexity of the models, and their demands on computer time. None of these concerns are as pressing today as in the mid-1970s when the technique was first seriously explored. Nevertheless, Balant and Gex-Fabry (1990) conclude their relatively recent review with the words "more work is needed before this approach can be considered as validated and proposed as a routine tool during drug development." Note that at the time those words were written, the method had been in existence for over 2 decades and had been explored with dozens of drugs and toxic substances. In deference to these authors, it must be conceded that validation in humans is indeed very difficult — but that is precisely why good, detailed models are needed. Where experiments cannot be readily performed, a reliable technique for predicting human drug distribution will be particularly useful. If, as has been repeatedly shown, a model created with data from mice or rats can predict pharmacokinetics in dogs or monkeys and those predictions can then be validated experimentally (which has now been done for many drugs), then model validation in humans becomes less of an issue.

It is the author's belief that the insight into drug distribution that can be obtained from PB-PK modeling amply justifies the additional investment in experimental work. Despite the uncertainties inherent in interspecies scaling, and the difficulties of validating human PB-PK models, even an approximately correct indication of human tissue distribution is of enormous value in clinical protocol design and in making drug development decisions. For these reasons, the author believes that interspecies scaling based upon PB-PK models will indeed become a routine part of drug development.

CHAPTER 4

Prediction of Drug Metabolism and Toxicity

I. THE GENERALIZED METABOLIC RULES OF JENNER AND TESTA

Considering the enormous number of foreign chemicals to which animal bodies may be exposed, and the thousands of cellular enzymes with which they could, in principle, interact, the science of drug metabolism may seem, at first glimpse, to be one of totally intractable complexity. Jenner and Testa (1980, 1981; see also Testa and Jenner, 1976), following guidelines laid down by many previous authors, showed, however, that the pathways of drug metabolism could be classified into a comprehensible scheme. The present, very sketchy overview is indebted chiefly to the book by Gibson and Skett (1986).

The reactions of drug metabolism are classified into three phases. Phase I reactions involve addition of a functional group (e.g., a hydroxyl), and phase II reactions are conjugations (e.g., glucuronidation). These conjugates are often excreted in the bile. Phase III metabolism involves the cleavage of phase II metabolites by the intestinal flora, which may be followed by reabsorption (enterohepatic circulation).

For many drugs the major route of phase I metabolism is by enzymes of the cytochrome P450 superfamily, which are found mainly in liver, but also in kidney, lung, and intestine. Reactions catalyzed by this system require molecular oxygen and NADPH, and include:

1. Oxidations, e.g., aromatic hydroxylation, aliphatic hydroxylation, epoxidation, N oxidation, and S oxidation. The aryl hydrocarbon hydroxylase activity responsible for oxidative activation of polycyclic carcinogenics such as benzpyrene has been identified as the P450 subtype CYP1A1; this form is primarily extrahepatic, occurring in lung, for example.
2. Dealkylation reactions, e.g., N dealkylation, O dealkylation, S dealkylation; in these reactions the alkyl group is lost as the corresponding aldehyde (a methyl group, for example, leaving as formaldehyde). These reactions are believed to proceed through a hydroxyalkyl intermediate.
3. Some secondary amines (e.g., amphetamine) are metabolized to release ammonia, leaving the corresponding ketone.

4. Phosphothionate oxidation, in which the phosphothionate sulfur is replaced by oxygen. This is an important metabolic reaction for phosphothionate insecticides, such as parathion.
5. Dehalogenation; halogenated aliphatic compounds may undergo oxidative dechlorination or debromination to yield the corresponding alcohol (or, in the case of dihalocompounds, such as halothane, the carboxylic acid).
6. The cytochrome P450 subtype CYP2E1 also catalyzes oxidation of ethanol in liver microsomes, an inducible reaction that complements the constitutive alcohol dehydrogenase of the cytosol (Gibson and Skett, 1986).
7. Some other P450-catalyzed reactions are NADPH-dependent reductions and are inhibited by oxygen. This group of reactions includes azo and nitro reductions.

Human liver appears to contain at least 15 cytochrome P450 isoforms, falling into the three related families CYP1, CYP2, CYP3. The substrate specificities of this highly polymorphic group of enzymes are different, but overlapping. Birkett et al. (1993) state that even a single amino acid substitution may change substrate specificity. Wrighton et al. (1993) provide information on which subtypes are believed to oxidize particular drugs.

In addition to cytochrome P450, several other enzymes of liver microsomes and cytosol are involved in reactions of phase I metabolism. This group of enzymes includes alcohol dehydrogenase, aldehyde dehydrogenase, aldehyde oxidase, and xanthine oxidase. These are soluble enzymes. Monoamine oxidase converts amines to aldehydes; it is principally mitochondrial. Liver also contains a diamine oxidase. Another group of monooxgenases, found in liver microsomes, are the flavin-containing monooxygenases (FMOs), which catalyze direct oxidation on nitrogen, sulfur, and phosphorus. These enzymes require NADPH and molecular oxygen, but are flavoenzymes, rather than cytochrome-containing enzymes. There are at least five members of the FMO family, of which one is primarily fetal. The recent literature on FMOs has been reviewed by Ziegler (1993). Yet other non-P450-mediated oxidation reactions occurring in liver include aromatases and alkylhydrazine oxidase.

Several reductive reactions occur in liver microsomes that require NADPH, but are inhibited by oxygen; the P450-dependent azo- and nitroreductases were mentioned above. NADPH-cytochrome c-reductase can also catalyze these reactions. Other non-P450 reductive reactions are epoxide reductase and reductive dehalogenation reactions (including defluorination).

A number of hydrolytic reactions are considered to be part of phase I metabolism, including esterases, amidases, and reactions of carbamate hydrolysis. Finally, there is a group of miscellaneous reactions that do not fall into the above classification. Gibson and Skett (1986) give several examples of these.

In summary, the outcome of phase I metabolism is often to make drugs more hydrophilic, generally by adding a functional group such as hydroxyl, amino, thiol, or carboxylic acid. These primary metabolites are then often acted upon by enzymes of phase II metabolism.

The function of the reactions of phase II metabolism is detoxification, i.e., the conversion of xenobiotics or their phase I metabolites to water soluble products that are readily removed in urine or bile. UDP-Glucuronyltransferase is a family of microsomal enzymes found in all mammalian tissues; they transfer a glucuronide group to a phenol or alcoholic hydroxyl forming an O-β-glucuronide ether, to a -COOH group forming an O-β-glucuronide ester, to an amine (or amide or sulfonamide) to give an N-glucuronide, or to a thiol forming an S-glucuronide. O-Glucuronides that are excreted in bile may be cleaved by β-glucuronidase in the gut, and the original compound reabsorbed, giving an enterohepatic circulation. Nine human isoforms of UDP-glucuronyltransferase have been identified to date. Recent research on drug glucuronidation has been reviewed by Mulder (1992). Similarly to glucuronidation, but to a much lesser extent, UDP-glycosyltranferase reactions may couple drugs or their phase I metabolites to other sugars; e.g., if UDG-glucose is the sugar donor, a glucoside is formed; ribosides and xylosides are seen for some drugs.

Some drugs are metabolized by methylation reactions catalyzed by a family of methyltransferases that use S-adenosylmethionine (SAM) as the donor molecule. The methyltransferases are probably primarily enzymes for metabolism of endogenous metabolites (e.g., histamine), but they tend to have fairly broad specificity; e.g., histamine N-methyltransferase (which occurs in lung and skin tissue as well as in liver) can transfer a methyl group to a variety of imidazole drugs. Other enzymes can transfer a methyl group to oxygen (e.g., catechol O-methyl transferase) or sulfur (S-methyltransferase).

N-Acetyltransferases are found in the cytosol of a number of tissues and transfer an acetyl group from acetylcoenzyme-A mostly to aromatic amines and sulfonamides.

Sulfation is another conjugation pathway that is catalyzed by a family of enzymes, in this case cytosolic, the sulfotransferases. The donor molecule is 3'-phosphoadenosine-5'-phosphosulfate (PAPS). Phenols are the preferred acceptors, but alcohols, amines, and thiols can also accept a sulfate group. There are at least five sulfotransferases, some with low specificity and others specific, e.g., for a particular class of steroids. Many acceptors that are glucuronidated are also sulfated; in comparison with glucuronidation, sulfation is a high-affinity, low-capacity pathway.

Drugs that have carboxylic acid functions may be converted in liver or other cells into coenzyme-A derivatives, which then N-acylate the primary amino groups of several amino acids, chiefly glycine, but also glutamine, arginine, ornithine, and taurine. Gibson and Skett (1986) list the formation of hippuric acid (N-benzoylglycine) as a classical example of amino acid conjugation.

A detoxification scheme that is important for many electrophilic drugs is glutathione S conjugation. At least seven isoforms of glutathione S-transferase are located within the cytosol and microsomes of liver and many other tissues. The acceptor molecules for these reactions include nitro- and haloaromatics, nitro- and haloalkanes, epoxides, and sp_2 carbon centers. The resulting thioether tripeptide conjugates may be excreted in urine, but are often cleaved by gamma-glutamyltranspeptidase, and further modified ultimately to N-acetylcysteine thioether derivatives. A number of examples of drug metabolism mediated by glutathione S-transferases are discussed by Tew et al. (1993).

Of course, many drugs may be metabolized by more than one of these pathways, so that the number of possible metabolites for a complex drug may be in the hundreds. Often, a substitution reaction will introduce a chiral center, further increasing the complexities facing the drug development scientist who attempts to identify and isolate drug metabolites.

Typically, drug metabolism is studied first in rats, whose drug-metabolizing pathways are qualitatively similar to those of humans, though quantitative differences are often seen. The principal metabolites found in rat blood and urine have often been identified before a new drug is tested in humans. We would often like to predict whether the same metabolites will be found in humans as were seen in rats, and also how the overall rate of metabolism is likely to compare in the two species. Various approaches are used to predict what we are likely to see when the drug goes into humans.

II. *IN VITRO* METHODS FOR PREDICTION OF DRUG METABOLISM

We shall consider two general approaches to predicting the pathways and extent of metabolism of new drugs in humans. The first is to use *in vitro* preparations (e.g., suspensions of rat or human liver microsomes) to measure metabolic conversion experimentally, perhaps then incorporating the resulting data into a PB-PK model so that the *in vitro* data are used to modify an *in vivo* model. The second approach, which we shall consider below, is to adopt an entirely computational approach, using a rule- and knowledge-based expert system.

Birkett et al. (1993) have written an informative short review on the use of *in vitro* methods for prediction of human drug metabolism. As a minimum, it is usually possible to determine which enzymes are involved, and thus to predict some possible drug–drug interactions that result when two drugs compete for the same pathway of metabolism or when one drug induces an enzyme that metabolizes the other drug. Wrighton et al. (1993) give a more comprehensive review of the subject. These authors conclude that *in vitro* studies make it possible to (1) make comparisons between metabolite profiles for a new agent between experimental animals and humans, (2) identify any species-specific metabolites, and perhaps determine the animal species most appropriate for toxicology or ADME studies, (3) determine the enzymes responsible for the metabolic clearance of a new agent, and (4) use enzyme kinetics to model the *in vivo* situation. The following discussion is based primarily on these two sources. There are four basic *in vitro* techniques for study of hepatic metabolism: (1) subcellular fractions of liver — generally either microsomes or cytosol, (2) purified drug-metabolizing enzymes, (3) isolated hepatocytes, and (4) liver slices.

A. Subcellular Fractions of Liver

This is the most common technique for *in vitro* metabolism studies and most frequently uses the S9 fraction, i.e., liver homogenate from which nuclei and mito-chondria have been removed by centrifugation. Alternatively, washed microsomes or a microsome-free cytosol may be employed. Microsomes contain numerous enzymes of the cytochrome P450 superfamily (the hepatic mixed-function oxidases), and the flavin-containing monooxygenases, and some phase II enzymes (UDP-glucuronyltransferases and some glutathione S-transferases). Cytosol also contains glutathione S-transferases, sulfotransferases, methyltransferases, and N-acetyl trans-ferases. In order to see metabolism in an S9 fraction or washed microsome suspension it is necessary to add the required cofactors (NADPH for cytochrome P450 and UDP-glucuronic acid for UDP-glucuronyltransferases). Obviously, an assay is required that is capable of distinguishing between the parent compound and the metabolite (often an HPLC method is used). Often, preliminary work is conducted with rat liver microsomes, and following identification of metabolites, follow-up studies are conducted with human liver microsomes (which are commercially available). Wrighton et al. (1993) advocate the use of a panel of liver specimens from different individuals, each sample of which has been phenotypically characterized for its activity of the different metabolic enzymes. Rates of formation of metabolites in the individual samples can then be correlated with their content of the various enzymes, and the identity of the enzyme responsible for metabolism of the test drug can be inferred. Sometimes this information can be obtained by competition studies against drugs whose metabolism is known to be specifically by a particular enzyme.

B. Studies with Purified Drug-Metabolizing Enzymes

A number of cytochrome P450 subtypes and several phase II enzymes (including several glutathione S-transferases) have been cloned and expressed in E. coli and are readily available in purified form (Gonzalez and Korzekwa, 1995). This makes it possible to examine the susceptibility of a test drug to metabolism by these enzymes and to calculate kinetic parameters such as V_{max}, K_m, and intrinsic clearance (k_{cat}/K_m). Birkett et al. (1993) give a list of cloned human P450 isoforms, along with model substrates and specific inhibitors for the various types (in some cases, specific antibodies are also available for drug-metabolizing enzymes and may be used for Western blotting). These reagents may often be used to determine the pathway of metabolism of an investigational drug. Whether or not the drug affects the rate of conversion of a known substrate, whether an inducer that selectively upregulates a particular P450 isoform affects conversion of the test compound, will usually help to map out the metabolic pathway. For a recent survey of selective inhibitors of P450 isoforms and their use in in vitro studies of metabolic reactions, see Halpert (1995).

C. Isolated Hepatocytes

Methods for isolating viable rat hepatocytes have existed for many years and have more recently been applied to human hepatocytes. They may be preserved in suspension culture for many hours and may be cryopreserved. Hepatocytes may also be maintained as monolayer cultures, though such cultures lose differentiated hepatic characteristics, including drug-metabolizing enzymes, with time. Monolayer cultures, however, do still respond to inducers of P450 subtypes.

Transformed hepatic cell lines, i.e., hepatomas, have also been suggested as a convenient way of maintaining liverlike cells in continuous culture. The disadvantage of hepatomas is that they all, to a greater or lesser extent, have some loss of specialized liver function.

D. Liver Slices

The use of liver slices suspended in a nutrient solution is a technique that has been used in normal metabolism studies since the 1930s, and it continues to be a useful technique for study of drug metabolism. Liver slices retain, certainly to a greater extent than isolated hepatocytes, the structural integrity of the organ, and unlike isolated hepatocytes, they have not been exposed to collagenase, with its potential for causing damage to the cell membrane. However, liver slices are not so readily cryopreserved as isolated hepatocytes, which presents a limitation in their use for human studies.

E. *In Vitro–In Vivo* Correlations

Pang et al.(1993) discuss reasons why metabolism *in vivo* may differ from what is predicted from the *in vitro* results. Subtle effects of liver structure are lost in *in vitro* studies, e.g., the fact that oxygen tension, nutrient concentrations, and exposure to xenobiotics vary between the periportal region, the midzonal region, and the perihepatic venous region. Pang et al. point out that distinct active transport carriers are found in the different zones. These authors also consider the fact that metabolism of many drugs involves a series of sequential reactions, and there is evidence that the liver is spatially specialized to handle sequential reactions in a manner that is dependent upon the direction of flow. Thus, periportal cells may have higher levels of phase I enzymes, while perihepatic venous cells may have higher activity of phase II enzymes. Again, this subtlety will be lost in a homogeneous suspension of hepatocytes. Frequently, the suspending medium used for *in vitro* studies is not a good approximation to the plasma environment; e.g., the effect of protein binding on the effective drug concentration must be taken into account.

Rates of metabolism, multiple pathways of metabolism, and kinetic parameters for metabolic enzymes, measured *in vitro*, may be used as system parameters for a PB-PK model. If a drug is shown to be metabolized by two or more pathways, and the activity ratio of the different routes differs between rat and human, a PB-PK model may be constructed using rat data, validated with rat studies *in vivo*, and human parameters may then be substituted into the model. Similarly, if the kinetic parameters for a metabolic enzyme differ between rat and human, the human values may be substituted into a rat model.

III. EXPERT SYSTEMS FOR PREDICTION OF DRUG METABOLISM

The ideal expert system for predicting drug metabolism would start by examining the chemical structure of the test compound, would examine a data base containing all the known reactions of drug metabolism, and would list the enzymes of phase I metabolism for which the test compound would be a substrate. It would then do a QSAR analysis to determine how good a substrate the test compound would be for each of the phase I reactions. Next the system would go through the same exercise for each of the phase I metabolites and determine their susceptibility to enzymes of phase II metabolism.

No current expert system approaches this ideal, although it represents a goal that will eventually be approached. The concept of using an expert system for predicting xenobiotic metabolism was first conceived by Wipke et al. (1983) who described a program called XENO. A commercially available expert system, METABOLEX-PERT (Darvas, 1987; Darvas and Eker, 1990), has been widely available for several years and is in use at Agouron Pharmaceuticals and at many other pharmaceutical companies. It is based upon the generalized metabolic rules of Jenner and Testa (1980, 1981) and uses a data base that is claimed to be an expanded and updated version of that of Pfeifer and Borchert (1975 to 1983) and on Hawkins (1989). The data base contains, for various species, structures of metabolites formed and the percentage of that metabolite detected in the excreta. METABOLEXPERT allows the user to enter new reactions in the knowledge base, or an existing transformation reaction may be modified. The chemical structure of the test compound is entered either using a graphics routine or through MolNote (which allows structures to be described as a line of text) or from an existing ChemBase Molfile. The rule-based approach for suggesting probable metabolites is a retrosynthetic approach that is most useful for rapid identification of the sites on the molecule where metabolic transformations may occur. In addition, METABOLEXPERT can use the approach of reasoning by analogy on the basis of a representative collection of compounds with known metabolic trees.

The output of the program consists of a list of the predicted metabolites, with the excretion percentage for each compound in the species selected. The metabolites are listed according to their relative importance. The chemical structures of the predicted metabolites may be displayed and printed. Second-phase metabolites are also predicted and listed.

METABOLEXPERT has a number of limitations that, for the present, restrict its usefulness. Test compounds are limited to 115 heavy atoms (i.e., atoms other than hydrogen); this is not a serious limitation, as it should include most drugs up to about 1700 Da molecular weight. There are limitations in the usable data base size. In addition, there are annoying limitations in the user's ability to search the data base.

How accurate is METABOLEXPERT? It makes many erroneous predictions and fails to predict many actual transformations. Some users believe that the expert system approach to predicting drug metabolism is limited by the fact that even a single atom change in structure may open up, or eliminate, the potential for a totally different route of metabolism. Of course, this does not mean that a rule-based approach is in principle useless, rather that there are still many unknown, or incompletely known, rules, which seriously limits the practical usefulness of this approach today, whatever its potential may be for the future. At Agouron Pharmaceuticals, the PC version of METABOLEXPERT 9.2 is used, not to avoid the experimental work of looking for metabolites of new chemical entities, but to provide a first-approximation guide to what we should be looking for.

A new expert system, META, has been described for prediction of biodegradation of chemicals (Klopman, 1995). META can reportedly be trained to recognize structural features of compounds that may result in susceptibility to particular routes of metabolism in different species. META is used in conjunction with the MULTICASE program (see Appendix).

Molecular Design Ltd. (MDL) offers a product called Xenobiotic Metabolism. Unlike METABOLEXPERT, Xenobiotic Metabolism is not an expert system, it is a data base. Like METABOLEXPERT, however, it uses as its starting point the five volumes of *Biotransformation von Arzneimitteln* (Pfeifer and Borchert, 1975 to 1983) supplemented by literature from over 60 journals covering the period 1984 to 1992, with annual updates promised for 1993 and later. Xenobiotic Metabolism includes information on species, route of administration, excretion, analytical methods, parent compound class, and biological activity of the parent compound. Xenobiotic Metabolism can be searched by ISIS™ (MDL's integrated scientific information system) or by REACCS (a reaction indexing system). Thus, the user who needs to predict how a new chemical entity is likely to be metabolized can do substructure searches to display known metabolites of related compounds or fragments. Xenobiotic Metabolism, unlike METABOLEXPERT, thus provides no help with the rules of metabolism, but has the advantage that its data base is totally open and searchable.

IV. *IN VITRO* TOXICOLOGY

There are many parallels between the prediction of drug metabolism and the prediction of toxicity. In both cases, we cannot usually obtain early *in vivo* data from humans, but we can obtain *in vivo* data from rats or other laboratory animals, which often has excellent predictive value, but which raises questions about possible species differences. In both cases, it is often possible to supplement animal *in vivo* data with comparative animal and human *in vitro* data. In both cases, computational approaches have been explored that may supplement limited experimental data.

The best-known and most widely used *in vitro* toxicity test is a cellular cytotoxicity assay that is readily automated, can be run in 96-well microtitre plates, and is equally applicable to rodent or human cell lines. However, there is a growing catalog of *in vitro* test systems than can measure more specialized effects. The multicenter evaluation of *in vitro* cytotoxicity (MEIC) project organized by the Scandinavian Society for Cell Toxicology is a large collaboration for establishing and validating *in vitro* toxicity test methods. An early study from this collaboration tested ten MEIC compounds in five cytotoxicity assays, using two human cell lines, primary rat hepatocytes, and a mouse cell line, and showed that this *in vitro* panel could predict the human lethal dose as accurately as the mouse median lethal dose values (LD_{50}) could. Wang and Ohno (1995) used lactate dehydrogenase (LDH) assays to measure both cell growth inhibition (CGI) and cell killing (CK). The former was measured by total LDH in the cell layer, and the latter was measured by release of LDH into the growth medium. They considered five types of *in vitro* toxic action. Type I had similar concentrations for 50% CGI and CK, almost parallel slopes for CGI and CK, and CK of 70% or more when CGI was 100%. Type II gave little or no CK at low concentrations where CGI was apparent, i.e., there was a threshold concentration for CK; the authors distinguished between type II-1, where CK reached or approached 100% at high chemical concentrations, and type II-2, where CK did not approach 100% even at high concentrations. Type III dose–response curves were characterized by the fact that the compound gave negative CGI (i.e., growth stimu-lation) at low concentrations, and this class was again divided into type III-1 and III-2, according to whether the compound did or did not give a high level of cell kill at high concentrations. These characteristic dose–response curves are summa-rized in Figure 4.1. While the *in vivo* significance of this classification is not entirely obvious, some generalizations may be drawn: for compounds that are highly lethal to cells, the CK and CGI curves will be close together (type I), while some com-pounds may be much less cytotoxic even at growth-inhibitory concentrations (type II-2 and type III-2).

Several *in vitro* test systems have been devised that attempt to predict toxicity to particular organs or tissues. For a recent general review, see Rogiers et al. (1993). Rowles et al. (1995) discuss the use of SH-SY5Y human neuroblastoma cells to predict neurotoxicity. This transformed cell line expresses a number of differentiated neuronal characteristics and had previously been shown to be susceptible to known neurotoxicants such as pesticides and heavy metals. Rowles et al. (1995) explored end points other than cell death, pointing out that neurotoxicity is not necessarily related to cytotoxicity. As sample end points related to neurotransmitter function they selected inhibition of acetylcholinesterase and neurotoxic esterase and dopa-mine D_2 receptor density. Effects of five known neurotoxicants, of different mech-anisms, were studied: mipafox (an organophosphorus esterase inhibitor); 3,3'-imin-odiproprionitrile, IDPN (a neurotoxic solvent); lead acetate; 1-methyl-4-phenyl-1,2,3,6-tetrahydropyridine, MPTP (a drug of abuse); and 1-methyl-4-phenylpyridinium, MPP$^+$, the oxidized analogue of MPTP. After 24-h treatment, mipafox, MPTP, and MPP$^+$ gave a dose-dependent inhibition of acetylcholinesterase, and mipafox also inhibited neurotoxic esterase. IDPN and lead acetate did not give acetylcholinesterase inhibition. Lead acetate, MPTP, and MPP$^+$ gave minor decreases in dopamine D_2 receptor density. These three agents also showed toxicity in a test for general membrane integrity, the ability to retain the fluorescent dye 5-carboxylfluorescein diacetate (CFDA). All these effects were seen at noncytotoxic concentrations. With the exception of IDPN, each test compound affected a neural-specific end point at concentrations at least 100-fold less than concentrations that affected cell viability.

IDPN, a known neurotoxicant, had no effect on the two end points used in this study. This compound has been reported in other *in vitro* studies to damage neu-rofilament networks. This observation raises one of the objections to biochemical end points for *in vitro* toxicity assays — in principle, one may need an assay system for every possible mechanism of toxicity. Test systems of the kind discussed by Rowles et al. are useful for mechanistic studies, but not well suited to a general-purpose early screen for neurotoxicity. Possibly a small panel of biochemical tests could be used to screen for some of the more common mechanisms of neurotoxicity.

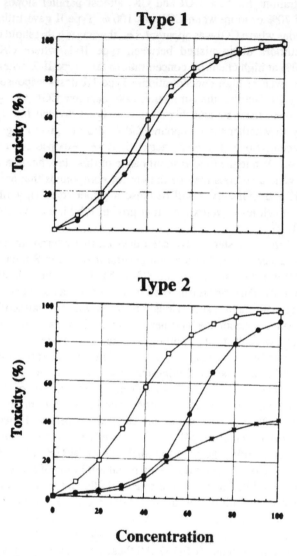

Figure 4.1 The five kinds of cytotoxicity dose–response curves. In each case, open squares show cell growth inhibition and the other symbols show cell killing. Upper panel: Type I; Middle panel: ● Type II-1, * Type II-2; Lower panel: ● Type III-1, * Type III-2.

An example of a functional assay for toxicity (in this case immunotoxicity) measured *in vitro* is seen in a study by Kobayashi et al. (1995) of effects of 12 MEIC compounds on natural killer (NK) cell function. Human NK cells may be isolated from peripheral blood mononuclear cells, and these cells are then cytotoxic to other cells, e.g., certain ^{51}Cr-labeled tumor cells. Eight of the twelve MEIC compounds tested positive in this assay. Activity in the NK-cell assay generally correlated with cytotoxicity to human cell cultures, though the levels that gave NK cell inhibition were about fourfold

Type 3

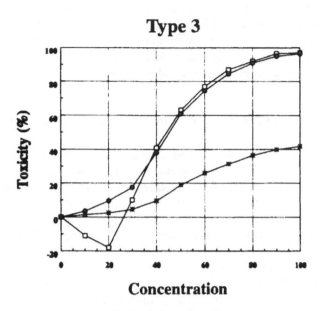

Figure 4.1 (continued)

lower than cytotoxic levels. While this was an interesting preliminary study, it did not attempt to validate the procedure with any *in vitro–in vivo* correlations.

Other *in vitro* tests for immunosuppression exist, such as the mixed lymphocyte reaction (MLR), a test for cytotoxic lymphocyte activity that is readily run with cells from a few milliliters of blood.

Advanced Tissue Sciences (La Jolla, California) grows cells from human foreskin on a synthetic polymer mesh to produce a skinlike material that has been used to test pharmaceuticals, cosmetics, and household chemicals. Results correlate well with skin damage measured *in vivo*; nevertheless, it must be remembered that this simple synthetic skin does not contain sweat glands, sebaceous glands, or hair follicles, and thus is an oversimplified representation of human skin.

A. How Are *In Vitro* Toxicity Data Used in Drug Development?

At present, *in vitro* toxicity assays are used mostly for a few specialized purposes. Cytotoxicity assays are used, for example, to test antimicrobial agents to make sure that an adequate safety margin exists between the MIC and the toxicity to host cells. Sometimes more specialized assays are used when a particular chemical series is known to cause a particular kind of toxicity. For example, the vinca alkaloid family of anticancer agents tend to cause peripheral neuropathy. Since vinca alkaloids act by binding to tubulin, and tubulin is vital for transport of nutrients along the axon, it was reasonable to assume that the neurotoxicity was a direct consequence of inhibition of axonal transport. *In vitro* measurements of inhibition of axonal transport correlated well with peripheral neuropathy in humans.

What we have not yet seen is *in vitro* toxicology assays becoming part of the routine workup of drugs for IND filing. It is likely that much of the animal toxicology currently done could be at least partly replaced by a panel of *in vitro* tests. Of course, this is unlikely to happen without some kind of signal from the FDA that *in vitro* toxicology is gaining acceptance for regulatory purposes.

V. EXPERT SYSTEMS FOR PREDICTION OF DRUG TOXICITY

As with drug metabolism, there are two general approaches to supplementing toxicology data from animals with predictions about what is likely to be seen in humans; the first, as discussed above, is experimental studies *in vitro*, and the second is the use of a data base with some computational rules about how to extrapolate conclusions from existing compounds to new compounds. As an example of a toxicology expert system, we shall consider the example of TOPKAT 3.0 (see Appendix). TOPKAT is an SAR-based procedure. It accepts the structure of the test compound either graphically or as a simplified molecular input line entry system (SMILES) string (Weiniger, 1988). Having assembled a data base, TOPKAT generates SAR parameters that can predict the required toxicity for new compounds. The SAR parameters include discrete descriptors, which identify functional groups present in a structure (e.g., aromatic nitro compound and ether ring linkage) and continuous descriptors, such as molecular weight, connectivity indices, partial atomic charges, kappa shape indices, polarizability, electron density, residual electronegativity, and topological symmetry indices.

TOPKAT 3.0 is available with the following prediction modules: rat oral LD_{50}, rat chronic LOAEL, rat maximum tolerated dose, mouse inhalation LC_{50}, rodent carcinogenicity, teratogenicity, Ames mutagenicity, developmental toxicity potential, rabbit skin irritancy (Draize), rabbit eye irritancy (Draize), fathead minnow LC_{50}, *Daphnia magna* EC_{50}, and aerobic biodegradability.

The predictive power of TOPKAT appears to depend upon whether the test compound is reasonably closely related to compounds that were present in the data base used for development of the prediction module. TOPKAT has been criticized for not distinguishing between positional isomers. Literature from Health Designs, Inc. (HDI Toxicology Newsletter, No. 16, March 1993) discusses TOPKAT evaluation of 100 compounds tested in the Ames test for mutagenicity by the National Toxicology Program. Of the 100 compounds, 12 were part of the data base from which the SAR module was developed, so were not used in calculation of predictive value. Sixteen chemicals were not adequately covered by the module data base; another 11 fell in the indeterminate probability range. Of the remaining 61 compounds, TOPKAT correctly predicted the Ames test result for 46 (75%), it gave a false-positive result for 8, and a false-negative result for 7. The discussion of this study goes on to show that if more stringent rules were used to judge the adequacy of coverage by the data base, and if estimates with a low confidence level were ignored, TOPKAT gave 44/48 correct predictions (92%) based on these more stringent criteria.

For compounds where the data base of the current modules does not provide adequate coverage or for biological end points where no TOPKAT module exists, Health Designs, Inc. sells a program, PROGNOSYS 1.0, for creating prediction modules. To design and validate new modules using this tool probably requires a higher level of expertise than evaluating new compounds using an existing module.

Other expert systems for predicting toxicity have been described, though this author has no experience with them. Lewis (1992) has reviewed the available programs. COMPACT (Lewis et al., 1990) uses a QSAR-based approach, factoring in biochemistry and cluster analysis. It was developed specifically to predict toxicity of compounds that are activated by the cytochrome P450 system (such as polycyclic hydrocarbons), and molecular shape parameters of polycyclic hydrocarbons (e.g., the ratio of area to depth^2) have been shown to predict for carcinogenicity. If area/depth^2 is plotted against delta E (electronic activation energy, an electronic parameter obtained from molecular orbital calculations), carcinogenic and noncarcinogenic compounds tend to fall into two distinct clusters. Similarly, plotting the ratio of molecular length to width against area/depth^2 gave clusters that were or were not substrates for cytochrome P450 type 3. Lewis (1992), one of COMPACT's inventors, gives a detailed description of it in his review.

DEREK (deduced estimation of risk from existing knowledge) is an expert system that recognizes molecular fragments known to be associated with various kinds of toxicity (Sanderson and Earnshaw, 1991). DEREK is currently restricted to molecules containing 64 or fewer atoms and provides only a qualitative estimate of toxicity. Similarly, HAZARDEXPERT is a fragment-based expert system for prediction of toxicity from the creators of METABOLEXPERT, and like the latter program, HAZARDEXPERT can predict metabolite formation and thus factor the metabolites into the overall toxic risk of a compound. HAZARDEXPERT, unlike DEREK, gives quantitative estimates of toxicity. It is available for PC or for VAX.

Lewis (1992) gives an interesting comparison of the ability of COMPACT and DEREK to predict activity in the Ames test for mutagenicity of 40 compounds tested by the National Toxicology Program. COMPACT found 16 true positives, 8 true negatives, 10 false positives, 5 false negatives, and 1 uncertain prediction, giving an overall 60% correct answers. DEREK gave 15 true positives, 12 true negatives, 6 false positives, and 5 false negatives and was unable to classify 2 compounds, for an overall 67.5% correct answers. COMPACT, however, is able to identify carcinogens that are not picked up by the Ames test (because they are not bacterial mutagens).

MULTICASE (Klopman, 1992) is a program that identifies molecular fragments with a high probability of being related to particular activities. It is designed to deal with "noncongeneric" sets of compounds that are not amenable to classical QSAR approaches. It requires, however, that the toxic effects of the various compounds be exerted by the same biochemical mechanism. The MULTICASE algorithm bears some resemblance to pattern recognition methods and principal component analysis. Unlike classical QSAR methods, MULTICASE selects its own descriptors automatically from a learning set composed of active and inactive molecules. Data bases of 30 to 50 compounds, distributed among active, marginally active, and inactive compounds, are required, and for optimal predictions larger data bases are desirable.

In conclusion, expert systems for prediction of drug toxicity are still evolving and currently have many limitations. In the opinion of Lewis (1992), no single program is preferable, and all have merit. He recommends combining several different methods and using these expert systems in conjunction with toxicity data bases such as Toxline, RTECS, and the U.S. Environmental Protection Agency (EPA) toxicity data base.

VI. QUANTITATIVE STRUCTURE–PHARMACOKINETIC RELATIONSHIPS (QSPR)

Quantitative structure–activity relationships (QSAR) is a multiple regression technique for quantitating the correlation between physicochemical properties of a series of compounds and some biological activity. The chemical features studied by QSAR include whole molecular properties such as charge or log P, local features (e.g., polarizability of a particular group), and substituent effects (e.g., fragment hydrophobicity indices). The biological property studied is typically binding to a receptor or inhibition of an enzyme, or sometimes an *in vivo* pharmacological end point (e.g., Dearden et al., 1989). However, any other "activity" may be studied just as easily, so the same approach can be used to correlate structural features with plasma half-life in a rat, or volume of distribution, or with a particular toxic activity. It is conventional to refer to studies of this kind as QSPR (QSTR), indicating that the activities of interest are pharmacokinetic (or toxicity) parameters. Clearly the use of QSPR is related to the expert systems for metabolism and toxicity prediction discussed in previous sections, except that commercial systems come with a preexisting data base and with the ability to do the structure–activity correlations automatically. Standard QSPR leaves much more work to the user. However, a brief survey of the literature will give some idea of what has been accomplished with QSPR. Two early studies in the field were that of Tong and Elien (1976) who described QSAR of biotransformation of barbiturates, tertiary amines, and substituted imidazoles, and Testa and Salvesen (1989) who discussed QSPR of amphetamines in humans. The early work was reviewed by Seydel and Schaper (1981), Lien (1981), Tichy (1985), and Hansch et al. (1989).

The PK of quaternary ammonium compounds were predicted by Neef et al. (1984), and Sawada et al. (1984) described the prediction of volumes of distribution of basic drugs in humans based upon animal data. Hinderling et al. (1984) described QSPR of beta-blockers in humans. Watari et al. (1988) used QSPR to predict hepatic first-pass metabolism and plasma levels following i.v. and oral administration of barbiturates. Cronin and Dearden (1989) correlated hydrophobicity and molecular connectivity with toxicities of chlorobenzenes to different species. QSPR of sulfonamides in the rat were described by Kaul and Ritschel (1990). More recent studies have combined classical QSAR methods with more qualitative approaches; e.g., human PK parameters of benzodiazepines were predicted using the graph theory method (Markin et al., 1988). Ghauri et al. (1992) combined QSAR with pattern recognition studies to predict metabolism of substituted benzoic acids in the rat. Franke et al. (1994) used combined factor analysis and QSAR to predict antibacterial and pharmacokinetic data from parallel biological measurements. Herman and Veng-Pedersen (1994) discussed the extension of QSPR to sets of compounds not confined to a single congeneric series, a problem that Hirono et al. (1994) approached using a fuzzy adaptive least-squares correlation method. Finally, Yamada et al. (1993) combined QSAR and QSPR of beta-blockers to predict therapeutic doses.

VII. BIOCHEMICAL SIMULATION IN TOXICITY PREDICTION

Where the biochemical mechanism of a particular toxicity is well understood, a number of questions about that toxic effect may be explored by kinetic simulation. This technique is discussed in detail in the section on pharmacodynamic modeling (Chapter 5), and an example will be given here that makes use of the author's BioCHIMICA model.

Antifolate drugs are strongly teratogenic, and because antifolates are in frequent use as anticancer, antibacterial, antiparasitic, antipsoriatic, and antiarthritic drugs, the questions of how to minimize their toxicity and whether their toxic effects may be separable from their therapeutic activity are of great practical importance. Jackson (1995) described studies with a biochemical model of 50 reactions of folate and nucleotide biochemistry in an attempt to predict some of the features of antifolate teratogenicity. It was assumed that this teratogenicity is a consequence of a cell attempting mitosis while it contains unrepaired DNA strand breaks, and that these breaks are caused by excision of misincorporated uracil. The reactions of the model are as shown in Figure 5.6, with the addition of various reactions of nucleotide biosynthesis and interconversion, including the formation of dUTP, resulting misincorporation of uracil units into DNA, and their excision. The simulation study considered four lipophilic antifolate drugs that inhibited thymidylate synthase (TS), dihydrofolate redutase (DHFR), or both. The properties of these four compounds are summarized in Table 4.1.

Table 4.1 Inhibition Kinetics of Quinazoline Antifolates

Inhibitor	K_i for inhibition of DHFR (nM)	K_i for inhibition of TS (nM)	IC_{50} against CCRF-CEM cells (μM)
AG199	5.0	360	3.4
AG225	10.0	210	1.0
AG337	>100.0	17	0.81
Trimetrexate	0.043	>1000	0.012

Note: K_i values were calculated assuming competitive interaction with the folate cofactor; IC_{50} values are for a 72-h treatment.

Other things being equal, an antimetabolite with a lower K_i value, i.e., tighter binding to its target enzyme, may be expected to be relatively more toxic than a less potent comparison compound, where other things in this context refers to *in vivo* properties such as tissue distribution or pharmacokinetic properties. Thus, estimating the relative toxic potency of two analogues of known inhibition kinetics may seem a straightforward exercise that should not require a complex kinetic model. However, many antimetabolites (which are often close analogues of normal metabolites) may interact with more than one enzyme, so that it is not always clear which is their primary target. In the present case, let us consider two of the compounds in Table 4.1. AG199 is a 360-nM inhibitor of TS and a 5-nM inhibitor of DHFR. AG225 is a 210-nM inhibitor of TS (1.7-fold more potent that AG199), but a 10-nM inhibitor of DHFR (twofold less potent than AG199). Which is the primary site of action for these compounds? It is not necessarily the enzyme for which the inhibitor has the lowest K_i value. Which compound should we expect to be most toxic? This can only be calculated using a model that is complex enough to factor in the enzyme kinetic parameters for DHFR and TS and to model the changes in substrate pools that occur in the presence of the inhibitors. This, in turn, necessitates modeling several other reactions. The predicted dose–response curves for activity of these compounds on human CCRF-CEM cells, as calculated by BioCHIMICA, are shown in Figure 4.2 (Jackson, 1995). AG199 was calculated to be slightly less potent than AG225, by a factor of 1.33 at the IC_{50}. The experimental data (for a 72-h exposure) showed that AG225 was, indeed, more cytotoxic than AG199. The dose–response curve for AG199 is slightly steeper than for AG225, so that the two curves become closer at high drug concentrations; this probably reflects the greater contribution of DHFR to the effect of AG199.

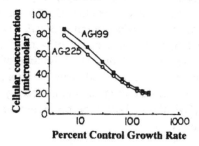

Figure 4.2   Dose–response curves for the bifunctional DHFR and TS inhibitors AG199 and AG225, as predicted by the BioCHIMICA model. (From Jackson, 1995. Reproduced with permission of Elsevier.)

Metabolic control analysis suggests that in the inhibited steady state, whichever of DHFR or TS has the larger control coefficient will be the primary target of the inhibitors. Calculated flux control coefficients in presence of an IC_{50} dose of AG225 were as follows: DHFR, 0.029; serine hydroxymethyltransferase, 0.001; and TS, 0.979. Clearly, the effect of AG225 upon TS is much more significant than its effect on DHFR. Another way of assessing the relative importance of the two targets is to use the BioCHIMICA program to (computationally) ignore the inhibition of either TS or of DHFR by AG225, and ask what would be the predicted effect of the resulting single-site inhibitors. When the effect of a pure TS inhibitor with K_i of 210 nM was calculated, the predicted IC_{50} was 26.3 μM, close to the value for the bifunctional inhibitor (24.5 μM). In contrast, for a pure DHFR inhibitor with K_i of 10 nM, the predicted IC_{50} was 97 μM. Thus, inhibition of TS is the dominant component of

AG225's cytostatic activity, though its effect on DHFR makes it about 7% more potent than it would otherwise be.

The shapes of dose–response curves for toxic substances have obvious implications for assessment of toxic risk, particularly the presence or absence of a threshold level. The BioCHIMICA model was used to predict dose-response curves for a pure TS inhibitor (AG337) and a pure DHFR inhibitor (trimetrexate). Figure 4.3 shows their calculated dose–response curves, normalized with respect to IC_{50} values. There is a proportionately much larger threshold region for the DHFR inhibitor. This is because the kinetics of the pathway are such that, through biochemical changes such as increases in the dihydrofolate pool, the system can absorb a much larger perturbation before the total flux starts to decrease. It appears that the toxic risk of very low drug concentrations is greater for inhibitors that act on enzymes with a high flux control coefficient.

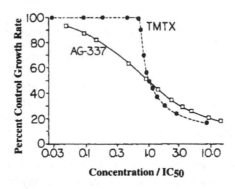

Figure 4.3 Relative dose–response curves for a pure DHFR inhibitor (trimetrexate) and a pure TS inhibitor (AG337). The concentrations are normalized with respect to their IC_{50} values. (From Jackson, 1995. Reproduced with permission of Elsevier.)

Another question about toxicity that we would like to explore by kinetic modeling is whether a toxic effect will be reversible or irreversible. When a biochemical system in a steady state is subjected to a small perturbation (e.g., with an inhibitor) the system will typically switch to a new steady state, though larger perturbations may make the system unable to achieve a steady state at all. If the inhibitor is removed before the cell suffers irreparable damage, it may be able to return to the original, uninhibited steady state. In some cases, however, a cell may be perturbed beyond a point of no return and be unable to get back to the original state after the inhibitor is removed. An example of a reversibility simulation is illustrated in Figure 4.4. This study simulated treatment of cells with the TS inhibitor AG337 at 10 μM for 60 min. The inhibitor was then removed. This particular model of folate metabolism contains 36 variable substrates, so may be considered to describe a system that moves within a 36-dimensional phase space. A perturbation of the system, as by addition or removal of an inhibitor, causes the system to describe a trajectory in phase space. Since we cannot print a 36-dimensional plot, Figure 4.4 plots the trajectory of the AG337-perturbed system with respect to [dUMP] and [dTTP]; after 60 min the system approached a steady state with high [dUMP] and low [dTTP]. Removal of

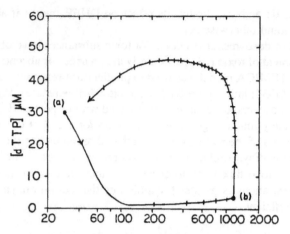

Figure 4.4   Perturbation of the biochemistry of human T-cells by (a) addition and (b) removal of the TS inhibitor AG337 (10 μ*M*), shown as a trajectory in two-dimensional phase space. The marks on the trajectory line indicate time in 10-min intervals. (From Jackson, 1995. Reproduced with permission of Elsevier.)

the AG337 then resulted in a new trajectory (not the reverse of the original trajectory) that, by 360 min, brought the system back close to the original uninhibited steady state. This concentration of this particular inhibitor, at least, results in a toxic perturbation that is fully reversible.

A final example of the use of kinetic simulation in prediction of toxicity is the prediction of combined effects of two or more toxic substances. In some cases, combinations of agents that, used alone, are innocuous may have a synergistic toxic effect. Since testing every agent in the presence of every other agent is impractical, this is an area where simulation may help by enabling experimental studies to focus on those combinations most likely to be problematic. Figure 4.5 shows results of simulations for the combinations of AG225 plus vidarabine triphosphate (ara-ATP) and AG225 + hydroxyurea, plotted in isobol form. The model predicted that the AG225 + ara-ATP combination should show moderate antagonism, but that the combination of AG225 plus hydroxyurea should be additive or slightly greater than additive.

To date, the potential of kinetic modeling as a tool for the prediction of toxicity has not been widely explored, perhaps because of the lack of user-friendly software. The availability of BioCHIMICA may help to rectify this deficiency.

VIII. PREDICTION OF LOG *P* AND PK$_a$

Ionization is a purely physicochemical process, and in principle it should be accurately predictable from the chemical structure of a compound. In practice, calculated pK$_a$ values are often fairly accurate. An example of a program that can estimate pK$_a$ values from the chemical structure of a compound is PKALC (see Appendix). The calculation of pKa utilizes the Taft and Hammett equations that relate additive free energy changes to the strength of ionization. The Dewar–Grisdale

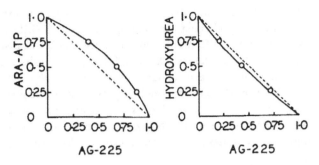

Figure 4.5 Isobol curves for drug–drug interactions between AG225 plus ara-ATP and AG225 plus hydroxyurea. Values on ordinate and abscissa show doses of the appropriate drug expressed as a fraction of their IC_{50} values. The broken lines show the position of additivity. (From Jackson, 1995. Reproduced with permission of Elsevier.)

method is used for compounds where the center and the substituents reside on different rings.

Lipophilicity is usually measured by medicinal chemists and pharmacologists in terms of the water–*n*-octanol partition coefficient (though other nonpolar solvents, e.g., propylene glycol dipelargonate, also have their adherents) and expressed as its logarithm, log P. Commercially available programs for measuring log P include CLOGP and SciLogP (based upon Bodor's equations) and PrologD, which calculated apparent log P taking into account the protonation state of ionizable groups. Pro-LogD allows the chemical structure of the compound of interest to be entered via a graphics routine, as a text string, or imported as a Molfile. The program calculates log P and pKa according to the methods of Hansch and Fujita, as elaborated by Hansch and Leo (1979), and further modified by Rekker. It then calculates the dependence of apparent log P on pK_a, finally producing a graph showing the dependence of apparent log P (log D) on pH. ProLogD is an open expert system that allows users to add new contributions to its data base.

Our experience of these methods has been that these programs generally order compounds correctly within a homologous series, but that absolute values may be off by one log unit or more. Nevertheless, these programs are useful tools, if supplemented with experimental log P determinations for selected compounds.

A. The Role of Large Data Bases in Drug Development

Most pharmaceutical companies as well as many government and academic research organizations have accumulated large chemical/biological data bases. Some of these are available commercially (e.g., the nonproprietary portion of the National Cancer Institute's data base may be obtained through Biosym). The *de facto* standard for such data base programs seems to be ORACLE (see Appendix). ORACLE is large, powerful (and expensive), and is run on large systems or networks, rather than on a PC. It is based upon IBM's Structured Query Language (SQL), a system for creating queries that can retrieve the required information. However, there is considerable portability between data base programs, so that a data base created by a researcher on a laboratory PC using, say, DBase can usually be imported or merged into the larger system. Many chemists and biologists use ISIS (MDL Information Systems, Inc., San Leandro, California) because it can store chemical structures, do substructure searches, and place data in a Microsoft Excel spreadsheet. ISIS can run on a network or an individual PC and can also communicate with ORACLE. Chemical/biological data bases of this kind are obviously extremely valuable in discovery research, for generating ideas for chemical leads, and for exploring structure–activity relationships, but as we have seen, the ability to sort compound lists by physical, chemical, or biological properties is also invaluable in development (e.g., for QSPR).

IX. PREDICTION OF ORAL BIOAVAILABILITY

Approaches to predicting oral drug availability from the physicochemical properties of the drug and from the physiological properties of the gastrointestinal (GI) tract have been reviewed by Oh et al. (1991), an excellent discussion to which the current treatment is greatly indebted. Factors mentioned by these authors include pK_a, solubility, dissolution rate, aqueous diffusivity, partition coefficient, chemical and enzymatic stability, intestinal pH, transit time, GI motility, endogenous substances such as bile salts, and exogenous substances such as food. Oh et al. distinguish between portal system availability and systemic bioavailability (which may also reflect first-pass metabolism in the liver). Even ignoring the effect of metabolism, the relationship between chemical structure and oral bioavailability is a very complex one, and currently it is not possible to make very accurate quantitative predictions of F% from chemical structure or physicochemical properties. Of course, there are a variety of heuristic rules relating F% to log P, solubility, or dissolution rate that can sometimes provide clues within a particular chemical series to making analogues that are more orally available. Dressman et al. (1985) defined absorption rankings (AR) in terms of n-octanol solubility, S_{oct}:

$$AR = \log (f \cdot S_{oct} \cdot V_L/EC_{50} \cdot BW) \tag{4.1}$$

where f is nonionized fraction at pH 6.5 (gut lumen pH), EC_{50} is some appropriate pharmacodynamic end point (which can often be measured *in vitro*), V_L is gut lumen volume, and BW is body weight. Dressman et al. (1985) defined absorption potential (AP) as

$$AP = \log (f \cdot P_{oct} \cdot S_w \cdot V_L/D) \tag{4.2}$$

In this equation, P_{oct} is the n-octanol–water partition coefficient, S_w is aqueous solubility, D is oral dose, and other terms are as defined for Equation 4.1.

Smith (1994) pointed out that the octanol solubility of many drugs can be predicted from ideal solution theory. He cited an equation of Yalkowsky et al. (1983) relating mole fractional solubility in n-octanol (X_{oct}) to melting point (mp):

$$\log X_{oct} = -0.01 \text{ mp} + 0.25 \tag{4.3}$$

This assumes that the entropy of fusion is essentially constant within a particular series, a reasonable assumption for rigid molecules. Smith (1994) showed a reasonable correlation ($r^2 = 0.81$) between the experimental F% and AR for a series of eight beta-blockers.

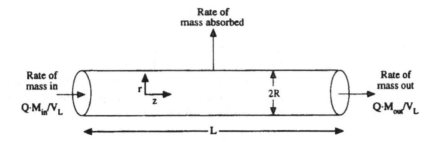

Figure 4.6 Macroscopic mass balance in the intestine. The rate of mass absorbed is the difference between rate of mass in and rate of mass out. V_L = lumen volume; Q = bulk flow; r and z are the radial and axial coordinates, respectively. (From Oh et al., 1991.)

The other kind of prediction that we would like to make is of F% in humans based upon experimental data in rodents (usually rats). Figure 4.6 illustrates a way of approaching this subject; the rate of intestinal drug absorption is

$$dM/dt = 2\pi R \int_0^L P_{eff} \cdot C_L(z)\,dz \tag{4.4}$$

where P_{eff} is the effective permeability (Amidon, 1983), a measure of the penetrating ability for a particular drug, and $C_L(z)$ is the local drug concentration at point z. For a compound that is absorbed passively, Oh et al. (1991) derived the following relationship:

$$F = 2A_n \int_0^1 C^* dz^* \tag{4.5}$$

where F is fraction absorbed, C^* is a dimensionless concentration variable, indicating concentration as a fraction of the initial concentration, and z^* is the axial coordinate expressed as a fraction of the total length of the intestine. A_n is a physiological parameter termed the absorption number; it is the ratio of radial mass transfer rate to axial convective flow rate, and it must be determined by experiments that measure intestinal wall permeability from single-pass perfusion experiments. A_n is affected not only by gut wall permeability, but also by the rate of bulk fluid flow in the intestine. The evaluation of the integral in Equation 4.5 requires a knowledge of how drug concentration changes throughout the length of the intestine (Oh et al., 1991). A_n is related to P_{eff} by:

$$A_n = \pi RL/Q \cdot P_{eff} \tag{4.6}$$

Note that the first term in Equation 4.6 contains anatomical parameters only, so may be used for interspecies scaling predictions.

If a drug is sufficiently soluble (or the dose is sufficiently low) that the drug content in the gut lumen is always in solution, then Oh et al. (1991) showed that:

$$F = 1 - \exp(-2A_n) \tag{4.7}$$

i.e., the fraction dose absorbed is exponentially related to A_n.

Another dimensionless parameter that Amidon et al. (1988) employ in prediction of oral absorption is the dose number (D_o), which is defined as the ratio of initial concentration to the solubility of a compound. For the situation where the concentration of drug leaving the gut exceeds its solubility:

$$F = 2A_n/D_o \tag{4.8}$$

(Oh et al., 1991). Thus, in this particular situation (applying to rather insoluble drugs) the fraction of dose absorbed decreases as dose number increases.

Another experimental technique that attempts to predict ability of compounds to cross the wall of gut epithelial cells uses a two-chambered vessel in which the chambers are separated by a porous membrane supporting a confluent culture of cells. Compound is then applied to the chamber corresponding to the enteric surface of the culture, and the ability of the compound to appear in the other chamber, corresponding to the basement membrane surface of the cells, is measured. The cell line generally used for this purpose is Caco-2 (Chantret et al., 1988), since it spontaneously differentiates in confluent cultures, establishes tight junctions between the cells, and expresses polarity and many other differentiated features. Measurements of permeability made with Caco-2 cells may be related to P_{eff} and substituted into Equation 4.6 to derive an estimate of A_n; this in turn may be used with Equation 4.7 or 4.8 to determine F.

Ong et al. (1995) described the use of immobilized artificial membranes (IAMs) to predict oral drug absorption. These membranes consist of cell membrane phospholipids covalently bonded to a solid chromatographic support. Solute capacity factors k_{IAM}' measured on IAM columns correlated with solute equilibrium partition coefficients K_m' measured with liposomes, indicating that solute partitioning between the IAM bonded phase and the aqueous mobile phase was similar to the solute partitioning between liposomes and aqueous phase. Both systems are believed to emulate fluid cell membranes, and the authors claim that IAM chromatography is experimentally more convenient. The binding of solutes to IAMs was independent of the chemical structure of the phosphatidylcholine ligand (based upon measurements with three such ligands), suggesting that the partitioning was a bulk phase property. Log P (n-octanol–water) did not correlate with k_{IAM}' except in homologous series of hydrophobic solutes (Pidgeon et al., 1995).

In summary, our understanding of oral drug absorption is at present encapsulated in a number of heuristics or simple empirical relationships concerning solubility, log P, and other physicochemical properties, supplemented with sophisticated physiological measurements on a small number of drugs that have provided insight into the process of drug absorption, but which are probably impractical as part of the normal preclinical workup for investigational drugs. The agenda of the current work is that heuristic approaches are increasingly likely to give way to knowledge-based and modeling approaches. Future methods of estimating oral absorption will probably relate *in vitro* data obtained with Caco-2 cells to a PB-PK model that includes a detailed description of the intestinal compartments.

CHAPTER 5

Pharmacodynamic Modeling

While pharmacokinetics is the study of drug concentrations in particular compartments as a function of time, pharmacodynamics is concerned with drug **effects**. Because most drugs are small molecules that act through binding to an enzyme or to a macromolecular receptor, pharmacodynamics is closely related to the science of enzyme kinetics, which relates the rate of an enzyme-catalyzed reaction to the concentrations of its reactants (substrates, activators, inhibitors, and the enzyme itself). A drug effect may be as simple as inhibition of an enzyme, or it may involve interaction with a more complex system, as when, for example, an ion-channel blocker causes a lowering of blood pressure. Pharmacodynamics is thus concerned with drug effects over a range of levels of complexity. The general scope of the subject is outlined in a chapter of *Goodman and Gilman* (Gilman et al., 1980). These authors emphasize three basic tenets of pharmacodynamics: the central position of the dose–response relationship in pharmacology, the importance of understanding a drug's mechanism of action, and the concept of selectivity. As in pharmacokinetics, pharmacodynamic parameters may be scaled across species, either by allometry or by physiological modeling (Mordenti and Green, 1991), so that we may attempt to use preclinical pharmacodynamic data to predict the response of humans to a developmental drug. A recent work that reviews the role of pharmacodynamics in clinical drug development is Yacobi et al. (1993).

Pharmacodynamic modeling, in practical terms, is usually an attempt to describe the dependence of a pharmacological effect, in a complex system, as a function of administered drug dose. For example, when the anticancer agent methotrexate inhibits its target enzyme dihydrofolate reductase, the inhibition of enzymatic activity is described very precisely by the Morrison equation (Equation 5.12). However, the pharmacological end point of interest is the inhibition of tumor growth, either *in vitro*, at the cellular level, or *in vivo*, in a tumor-bearing animal. These effects are usually much more complex functions of drug concentration than enzyme kinetic relationships. They are often described by empirically fitted curves, though, as we shall see, attempts can be made to derive mechanistic models for these complex processes. A recent discussion of the clinical pharmacodynamics of antiarthritic drugs referred to such behavioral end points as grip strength and walking time. These measurements are not only likely to be a very complex function of drug dose, but they also involve many complexities of quantitation (Furst, 1993). In such cases, it is probably unrealistic to attempt to develop a mechanistic model, but dose–response relationships may still be described empirically, so that even these behavioral end points can be related in a quantitative way to the administered drug dose.

Depending upon the nature of the end point, a pharmacodynamic effect may be separated in time from the drug concentration to which it is a response. For example, a drug effect may be related, not to the instantaneous drug concentration, but to the accumulated effect over a certain period of time. In some such cases these effects may be modeled mechanistically (e.g., for some anticancer drugs, cell kill may be related to progressive accumulation of a toxic metabolite); in other cases the end point may be empirically related to the drug's area under the curve (AUC).

Pharmacodynamic modeling, then, is an attempt to explain and predict these often complicated dose–response relationships in terms of the kinetics of multienzyme systems, of cell kinetics, or of whole animal biology, to describe them as functions of drug dose and of time, and to relate them to the anatomical compartment in which the drug is acting. For the purposes of the present discussion, we shall focus on pharmacodynamics as the study of the nature, extent, and consequences of the interaction of a drug molecule with its receptor. Models of drug effects on cytokinetics, i.e., the growth kinetics of cells (whether normal animal cells, or tumor cells, or those of pathogens), are described in detail in the next chapter. Strictly speaking, cytokinetic effects caused by a drug may be considered as an aspect of its pharmacodynamics. However, because cytokinetics has developed as an independent discipline, it is convenient to consider cytokinetic models separately.

I. KINETICS OF ENZYME AND RECEPTOR INTERACTIONS WITH LIGANDS

A very brief survey of the kinetics of protein–ligand interactions is necessary here, as pharmacodynamics is inevitably closely linked with biochemical kinetics. The present survey confines itself to discussing the most useful equations; for further information and derivations, readers are referred to the standard texts (Segal, 1975; Webb, 1963).

A. The Hill Equation

The Hill equation is an expression of the law of mass action:

$$v = [V_{max} \, S^n]/(S^n + K_m) \tag{5.1}$$

where v is reaction velocity, V_{max} is maximal velocity (when enzyme is saturated with substrate), S is substrate concentration, and K_m is a constant. The form of this equation implies that the reaction involves n ligand molecules. This situation may arise, for example, if an enzyme has two identical subunits (each of which binds substrate) and if the substrate binding is highly cooperative (i.e., if binding of the first substrate molecule greatly facilitates binding of the second). In such a case the value of n (the Hill coefficient) would probably be close to 2. Where the Hill coefficient is 1, the expression reduces to the familiar Michaelis–Menten equation. The constant K_m is known as the Michaelis constant and corresponds to the substrate concentration at which v is half the maximal rate.

The binding of an agonist to a receptor may be described by an analogous equation:

$$f = RA/R_t = A^n/(A^n + K_d) \tag{5.2}$$

In this case, f is the fraction of the receptor molecules that have bound ligand, R_t is the total concentration of receptors, A is the concentration of ligand (agonist), and RA is the concentration of receptors that have bound agonist. K_d is the dissociation constant for the binding process.

B. The Scatchard Equation

If we set n = 1 and rearrange Equation 5.2, we obtain:

$$RA/A = -1/K_d \cdot RA + R_t/K_d \tag{5.3}$$

Thus, a plot of the bound-to-free ratio against the bound concentration (RA) is linear with a slope of $-1/K_d$ and intercept R_t/K_d. Tallarida and Murray (1987) described a computer program for using Scatchard analysis to estimate dissociation constants and total concentration of binding sites (R_t) from experimental binding data. This program is part of the PHARM/PCS package. The classical Scatchard equation makes the assumptions that ligand binding does not show cooperativity, and that there is a single homogeneous class of receptor sites. For equations that describe the more general situation, see Thakur (1991).

C. Noncompetitive Inhibition

A noncompetitive enzyme inhibitor decreases the effective concentration of active enzyme by a factor of

$$i = 1 - (v_i/V_{max}) = 1 - [1/(1 + I^n/K_i)] \tag{5.4}$$

where i is the fractional inhibition, v_i is the inhibited reaction velocity, I is the inhibitor concentration, n is the number of inhibitor molecules reacting with each enzyme molecule (the Hill coefficient), and K_i is the inhibition constant (dissociation constant). The apparent V_{max} is given by:

$$v_i = V_{max}/(1 + I^n/K_i) \tag{5.5}$$

v_i can then be used (in place of V_{max}) in Equation 5.1 to obtain the reaction velocity. Note that for noncompetitive inhibition the fractional inhibition is independent of the substrate concentration.

D. Competitive Inhibition

Kinetically, the difference between noncompetitive and competitive inhibitors is that while noncompetitive inhibitors change the apparent value of V_{max}, competitive inhibitors change the apparent value of K_m. In practice, some compounds are found to have both effects to a greater or lesser degree, and such mixed inhibitors can be described by a $K_{i,slope}$, which defines the competitive effect, and a $K_{i,intercept}$, which describes the noncompetitive contribution to inhibition (note that many authors describe mixed inhibitors as noncompetitive). Structurally, competitive inhibitors are often substrate analogues, and mechanistically, they are believed to bind to the enzyme at the substrate site. The effect of a competitive inhibitor is given by the equation:

$$i = 1 - [(1 + S/K_m)/(1 + S/K_m + I/K_i)] \tag{5.6}$$

and $K_{m,app.}$, the apparent Michaelis constant in the presence of inhibitor is

$$K_{m,app.} = K_m \cdot (1 + I/K_i) \tag{5.7}$$

E. Ordered and Random Mechanisms

For enzymes that catalyze a reaction involving two or more reactants, a number of reaction mechanisms are possible. A distinction may be drawn between those enzymes that are indifferent to the order of addition of reactants and enzymes for which the reactants must bind in a particular order. Consider a reaction where substrates A and B react in the presence of enzyme E. Where one substrate (A) must bind first (so that substrate B then binds to the E–A complex), the kinetic mechanism is described as ordered sequential, and the rate equation is

$$v = V_{max}AB/(K_{ia}K_b + K_bA + K_aB + AB) \tag{5.8}$$

where K_a and K_b are the Michaelis constants for binding of A and B, respectively, and K_{ia} is the inhibition constant for inhibition of the back-reaction by A.

If the substrates are capable of binding in either order (i.e., $E + A \rightarrow EA + B \rightarrow EAB$, or $E + B \rightarrow EB + A \rightarrow EAB$), then the kinetic mechanism is random. If the form of the enzyme that binds substrates is in rapid equilibrium with the form of enzyme that is produced following the conversion of substrates to products, then $K_a = K_{ia}$. The mechanism is then described as rapid equilibrium random, and the rate equation reduces to

$$v = V_{max}AB/(A + K_a)/(B + K_b) \tag{5.9}$$

For some enzymes the first substrate binds, and its product is released (leaving a modified enzyme) before the second substrate can bind. This kind of mechanism is described as a "ping-pong". A common example of ping-pong reactions is found in phosphokinase enzymes, where the first substrate (ATP) binds and the first product (ADP) is released, leaving a phosphorylated enzyme; this then reacts with the phosphate-acceptor substrate, regenerating the original enzyme form. For a ping-pong mechanism, $K_{ia} = 0$, so Equation 5.8 is reduced to

$$v = V_{max}AB/(K_bA + K_aB + AB) \qquad (5.10)$$

The kinetic mechanism of a particular reaction may have an important influence on the regulation of metabolic pathways and on the effective use of inhibitors; e.g., examples are known of inhibitors of ordered sequential reactions that are analogues of the second substrate, so that their binding is sensitive to the cellular concentration of the first substrate.

F. The Monod Equation

This equation describes the kinetics of a reaction where (1) a substrate binds cooperatively to a multisubunit enzyme, so that the Hill coefficient is >1, and (2) an effector (inhibitor or activator) changes the affinity of the enzyme for its substrate with (3) a change in the Hill coefficient. An activator changes the substrate saturation curve to a more hyperbolic shape, and an inhibitor changes it to a more sigmoidal shape:

$$v = \frac{V_{max} \cdot S/K_m \ (1+S/K_m)^{n-1}}{L \cdot \dfrac{(1+I/K_i)^n}{(1+A/K_a)^n} + (1+S/K_m)^n} \qquad (5.11)$$

where A is an activator, K_a is its dissociation constant, L is an allosteric constant, and the other parameters are as previously defined.

G. Tight-Binding Inhibitors: The Morrison Equation

The rate equations discussed so far make a number of simplifying assumptions. One such assumption, reasonable in most situations, is that inhibitors (and substrates) are generally present at much higher molar concentrations than the enzymes to which they bind. If this is the case, then the total concentration of ligand will not be appreciably depleted by binding to the enzyme, so that total ligand concentration (which in an experimental study is usually known) and free ligand (generally unknown) can be treated as identical, which greatly simplifies the rate equations. Thus, when we study a noncompetitive inhibitor with a K_i value of $1 \times 10^{-6} \, M$ reacting with an enzyme that is present at $1 \times 10^{-8} \, M$, at 50% inhibition, the free inhibitor concentration will differ from the total inhibitor concentration by only

0.5%. However, for tight-binding inhibitors with very low K_i values, this assumption is not valid. What is meant by tight-binding in this context depends upon the concentration of the target enzyme; however, assuming that many enzymes are present in the cell at concentrations of around $1 \times 10^{-8}\,M$, any inhibitor with a K_i value of less than this will be likely to show significant deviations from Michaelis–Menten kinetics.

The kinetics of tight-binding inhibitors are described by an equation modified from a study by Morrison (1969):

$$v_i = v_o/2E_t \cdot \{(E_t - K_i - I) + [(Et - K_i - I)^2 + 4K_iE_t]^{1/2}\} \tag{5.12}$$

where v_i is the inhibited velocity, v_o is the uninhibited rate at the same substrate concentration, I is molar inhibitor concentration, E_t is molar enzyme concentration, and K_i is the inhibitor dissociation constant for a noncompetitive inhibitor. For a competitive inhibitor, K_i in Equation 5.12 is replaced by the apparent K_i, which equals $K_i (1 + S/K_m)$, S being the competing substrate. This is an important and useful equation, because many drugs function as tight-binding inhibitors.

H. The Relationship Between K_i and IC_{50}

For many studies in biochemical pharmacology, inhibitors are compared and data are expressed as the 50% inhibitory concentration, or IC_{50}. While IC_{50} is a convenient raw datum point, it is not a kinetic constant, because its value will usually depend upon the reactant concentrations. The exception is for a pure noncompetitive inhibitor, which is not tight binding. In this case $IC_{50} = K_i$. For tight-binding inhibitors, the relationship between IC_{50} and K_i was shown by Cha (1975) to be as follows:

Noncompetitive

$$IC_{50} = E_t/2 + K_i \tag{5.13}$$

Mixed competitive/noncompetitive

$$IC_{50} = E_t/2 + \frac{(S + K_m)}{\dfrac{K_m}{K_{is}} + \dfrac{S}{K_{ii}}} \tag{5.14}$$

Competitive

$$IC_{50} = (E_t/2 + K_i) + \frac{K_iS}{K_m} \tag{5.15}$$

In each case, the relationship between K_i and IC_{50} for non-tight-binding inhibitors may be obtained by setting $E_t = 0$.

II. DIFFERENCES BETWEEN INHIBITION KINETICS
IN THE TEST TUBE AND IN THE CELL

Enzyme kineticists generally study the properties of enzymes in dilute aqueous solution. The kinetic parameters that were derived in this way often provide a surprisingly good description of the behavior of the enzyme in the much more complex environment of the cell. Nevertheless, one should remember that the cellular environment may differ in ways that can affect the kinetics of enzyme systems. Within a subcellular compartment, enzymes may be present at very high local concentrations. If, as sometimes happens, the concentration of an enzyme is greater than the concentration of its substrate, the usual assumptions of steady-state enzyme kinetics no longer apply. For example, substrate may become saturated with enzyme. In such conditions, the rate equation is

$$v = V_{max} \cdot S/(E_t + K_m) \tag{5.16}$$

In the cell membrane, and in many of the subcellular organelles, it is likely that a number of enzymes exist in membrane-bound form. Where the catalytic site is a separate domain that extends into the aqueous phase, the kinetics are probably not greatly affected by the membrane association; for example, many cell surface receptors consist of an extracellular ligand-binding domain, a lipophilic membrane-spanning domain, and an intracellular catalytic domain (often a protein kinase), which is activated by ligand binding. The catalytic domain usually contains an ATP-binding site that appears to follow conventional aqueous kinetics. In other cases, especially where an enzyme exists entirely in a hydrophobic compartment and acts upon a lipid soluble substrate, it seems probable that the kinetics may differ markedly from what would be seen in an aqueous preparation (which, for such a system, would not be easy to study, although kinetic measurements have been made of detergent-solubilized membrane enzymes). We know very little of the kinetics of such systems.

In some cellular compartments, it is likely that enzyme may be so concentrated that it exists in the solid state, or to say the same thing in another way, some proteins may serve both a structural and an enzymatic role. Again, not too much is known of the kinetics of such systems. However, we do know that enzyme crystals usually retain full enzymatic activity, and perhaps this is not so surprising when we remember that enzyme crystals are typically 60 to 70% water. Thus, hydrated solid-state enzymes are probably best treated as an extreme case of high local concentrations. As such, it is quite likely that they may exist in the state mentioned above, where substrate becomes saturated with enzyme, and thus follow Equation 5.16, rather than classical Michaelis–Menten kinetics.

A. Compartmentation and Channeling

It is a common observation that the measured rate of a reaction in an intact cell may appear to be faster than can be accounted for on the basis of the enzyme activity as measured in a cell-free system and the known amount of substrate in the cell. However, in such cases the calculated rate is usually based on the assumption that substrate is evenly distributed throughout the cell's water space. Probably the usual explanation of this disparity is that the substrate is not evenly distributed, but confined to a small region of the cell, i.e., a subcellular compartment. The BioCHIMICA biochemical modeling program, which is discussed below, can simulate compartmentation. However, obtaining experimental measurements of metabolite concentrations is usually difficult, though for the compounds that are stable under the conditions of subcellular fractionation, it can sometimes be done. In the case of compounds that are not stable under these conditions, e.g., ATP, the concentration in a particular compartment can sometimes be deduced from the ratio of other (stable) molecules with which ADP and ATP may exist at or close to equilibrium.

There are other examples of a rate process in an intact cell being faster than can be accounted for on the basis of the individual reaction kinetics, where compartmentation cannot be invoked. In such cases it has been argued that the consecutive enzymes in a pathway may be organized such that the product of the first reaction (which is the substrate for the second reaction) has a high probability of binding to the second enzyme before it has the opportunity of diffusing off into free solution. This means that the local concentration of this compound is much higher than if it were evenly distributed throughout the cell. This situation can be modeled (at least approximately) by BioCHIMICA if we assume that the area in the vicinity of the second enzyme forms a separate, small compartment. Again, collecting good experimental data to fit the model to is very difficult. Most of the evidence for this kind of channeling comes from isotope dilution studies.

A rather special form of intramolecular channeling may occur in the situation where a single polypeptide chain contains two (or more) catalytic domains, and the product of one domain acts as the substrate for another. An example of this situation is the bifunctional dihydrofolate reductase–thymidylate synthase of protozoa. In this complex, dihydrofolate, a product of the thymidylate synthase reaction, acts as a substrate of dihydrofolate reductase. Knighton et al. (1994) studied the crystal structure of this bifunctional enzyme and showed that although no topological channel was present, a chain of positively charged residues extended from one domain to the other, which are presumed to maintain a high local concentration of the negatively charged dihydrofolate and herd it in the right direction.

To summarize the conclusions of this section, although most enzyme kinetics is studied in homogeneous, dilute, aqueous solution, and although the intracellular kinetics of many enzymes appear to be well described by this model, in some cases enzymes and their substrates may exist in heterogeneous, concentrated, or nonaqueous conditions or compartmented in such a way that the assumptions that we usually make when modeling multienzyme systems may not apply. If we are to use our computational tools to understand the kinetics of pathways in whole cells, we may need to extend the study of enzyme kinetics in ways that more closely reproduce the intracellular environment.

III. CELLULAR PHARMACOKINETICS

Physiologically based pharmacokinetics represents the accumulation of drug in a particular organ or tissue as being a function of blood flow to that tissue and of a partition function that describes how the drug will distribute between the blood and that tissue. The partition function is estimated either from ADME experiments *in vivo* or from *in vitro* studies with tissue homogenates. Sometimes it is possible to explain the affinity of a drug for a particular cell type in much greater detail. For example, entry of the anionic molecule methotrexate into cells is known to be facilitated by either of two transport mechanisms. Once inside the cell, methotrexate is converted through the enzyme folylpolyglutamate synthetase (FPGS) into long-chain polyglutamates, which, unlike free methotrexate, cannot readily efflux from cells. A polyglutamate hydrolase can convert the polyglutamates back to free methotrexate. The degree of accumulation of methotrexate in a particular cell type is highly dependent upon the activity of the two membrane carriers, of FPGS, and of the hydrolase. Most purine or pyrimidine nucleoside drugs, although electrically neutral, enter cells much more rapidly in the presence of a nucleoside transporter protein. Inside the cell, they are converted through the action of a sequence of kinases to mono-, di-, and triphosphate nucleotide forms. In some cases the triphosphates may be substrates for RNA or DNA polymerases, in which case the drug may be incorporated into nucleic acids. The nucleotide forms may be converted back to the nucleoside through the action of nucleotidases. Analogues of adenosine and cytidine may be subject to deamination within cells. Again, the degree of accumulation and retention of these nucleoside drugs in different cell types is highly dependent upon the activities of the nucleoside transport systems and on the activities of the cellular drug-metabolizing enzymes. The reactions of cellular uptake, activation, and turnover of a drug-constitute its **cellular pharmacokinetics.** A study of the cellular pharmacokinetics of a drug in different cell types will often present clues to understanding its selectivity. Clearly, the ability to model cellular pharmacokinetics provides a way to make predictions about probable drug activity, or lack of it, in particular tissue types or to estimate possible human activity from preclinical data. Some immunosuppressives are selective drugs because the cellular pharmacokinetics in lymphoblasts differs from the cellular pharmacokinetics in lymphoid cells. For some antitumor agents, the antitumor spectrum can be at least partially explained in terms of differential cellular accumulation or retention between different tumor types.

While we have a reasonably detailed understanding of the cellular pharmacokinetics of such drugs as methotrexate or cytarabine, which have been studied for decades, and it has been possible to construct pharmacokinetic–pharmacodynamic models (e.g., Morrison et al. 1975), for the typical development candidate we do not have much of this kind of information available before the compound enters clinical trials. An exception was gemcitabine (2′,2′-difluorodeoxycytidine), whose cellular pharmacokinetics and toxicity were studied by Heinemann et al. (1988). These authors attempted to explain the observation that gemcitabine had activity against several solid tumors in mice, unlike the closely related compound cytarabine, whose activity in mice and humans is generally confined to leukemia. Heinemann et al. found that membrane permeation of gemcitabine was 65% more rapid than that of cytarabine, that deoxycytidine kinase had a higher affinity for gemcitabine than for cytarabine, and that the elimination of the gemcitabine active metabolite 2'-difluorodeoxycytidine-5'-triphosphate (dFdCTP) was much slower than that of cytarabine's active metabolite ara-CTP. These factors resulted in cellular concentrations of dFdCTP that were up to 20-fold greater than cellular ara-CTP concentrations after treatment of cells with equimolar concentrations of the parent drug.

While it may never be feasible to complete a detailed evaluation of cellular pharmacokinetics on every compound before it enters clinical trials, because of the time and expense involved, there are a number of simple experiments that can enormously increase our understanding of how a compound is handled, and this information may exert a major influence on drug development decisions. Often these experiments take advantage of mutant cell lines. Consider the issue of whether a drug is actively transported. To measure transport directly requires either a very sensitive assay for small amounts of intracellular drug (which may be possible if the compound is highly fluorescent) or a supply of radiolabeled drug. However, if an appropriate mutant is available, all that is necessary is a comparison of the effects of the compound in normal and mutant cell lines. For example, many dihydrofolate reductase inhibitors, like methotrexate, enter cells through the reduced folate carrier. Cell lines are available, selected for resistance to methotrexate, in which the activity of this carrier is reduced to about 1% of the normal level. If we now test a novel antifolate drug in wild type cells and cells with the transport defect and see no difference in their IC_{50} values, we may be fairly certain that our new agent does not enter cells on that particular carrier. In the case of nucleoside drugs, there are cell lines available that have specific deletions of particular kinases, which can thus provide a rapid indication of whether or not that kinase is involved in activation of the compound under test. Another situation where cellular pharmacokinetic data is fairly easy to obtain is where a sensitive HPLC assay can be used to measure intracellular levels of drug and metabolites at different time points. This was the case with gemcitabine (Huang et al., 1991). If the site of action of the drug is known with some confidence and the active metabolite has been identified, the cellular pharmacokinetic data can be combined with pharmacodynamic modeling to produce a reasonable estimate of the expected drug effect. Of course, this requires a sophisticated pharmacodynamic model that represents the essential reactions of drug uptake, intracellular activation, and turnover. How this is done is discussed below, when the simulation of biochemical pathways is reviewed.

IV. EMPIRICAL AND MECHANISTIC MODELS

The relationship between drug concentration, determined by a pharmacokinetic model, and the drug effect may be as simple as a receptor or enzyme saturation function, or it may be a complex function involving a very large number of nonlinear processes. The pharmacodynamics may be modeled in two ways — mechanistically, where each step of the biological process is represented by an appropriate equation, or empirically, in which case the relationship is expressed by an equation that represents the experimental data accurately, but where the parameters are not simply related to physical components of the system. There are advantages and disadvantages to each approach. For example, the killing of a bacterial cell by an antibiotic may involve a complex chain of reactions, but the process can be measured experimentally very easily and the results expressed by a dose–response curve, even if the details of the compound's mechanism of action are unknown. Sometimes an empirical model is the only way to describe a complex effect quantitatively, and this will usually be the case with characterizing lethality, for example, where there is probably no drug for which a mechanistic model can generate an LD_{50}. Moreover, empirical models are easier than mechanistic models to characterize in terms of a small number of parameters. Many empirical pharmacodynamic models have been developed that relate the blood concentration of a drug to its therapeutic effect. Conolly and Andersen (1991) give a brief review of empirical pharmacodynamic models and point out a number of useful generalizations. A "link" equation is often used to describe the amount of drug in an "effect" compartment as a function of drug in the central compartment of the pharmacokinetic model. This link equation allows for a delay or hysteresis between the blood concentration of drug and the pharmacological response. In general, the model should be the simplest description that gives an adequate representation of the data, and the main criterion in selecting the mathematical structure of the effect model should be consistency with the pharmacokinetic and pharmacodynamic data. Thakur (1991) also discusses the relative merits of empirical and mechanistic models. Among the points he makes is that empirical models may be useful for purposes of interpolation, but that if we intend to extrapolate beyond the observable range, a mechanistic model is essential. In some cases, as Thakur points out, the distinction between empirical and mechanistic models may be blurred; for example, if the pharmacodynamic effect of interest is caused by receptor binding that follows the four-parameter logistic model, it may be possible to assign physical meaning to the four binding parameters, in which case the empirical model also has physical significance. For the most part, however, empirical models are used for their simplicity, for their amenability to statistical manipulation, or because the actual process being modeled is not sufficiently well understood to allow a mechanistic model to be constructed.

V. SHAPES AND SLOPES OF EMPIRICAL DOSE–RESPONSE CURVES

While the following list is not intended to be exhaustive, the ten empirical relationships will probably meet the needs of 99% of pharmacodynamic modeling applications. Users who need a larger selection of equations are referred to the TABLECURVE program (see Appendix), which lists (and fits) over 3300 equations.

A. Linear Dose–Response Curves

$$f = k_0 + k_1 \cdot A \tag{5.17}$$

where f is the effect (such as fraction of a receptor occupied), A is the drug concentration, k_1 is the first order rate constant, and k_0 is the baseline effect. Linear dose–response relationships are unusual in pharmacology, but crop up (to a first approximation, at least) in two kinds of situation: with cell-mediated cytotoxicity, such as tumor cell destruction by macrophages, and with very potent tight-binding inhibitors, or toxins such as ricin, where the dissociation constant for inhibitor is so low that the free inhibitor concentration at equilibrium is negligible. Linear dose–response relationships may be more common in other areas of biology; Brown and Rothery (1993) discuss an example from the area of nutrition.

B. Polynomial Dose–Response Curves

$$f = k_0 + k_1 \cdot A + k_2 \cdot A^2 + \ldots + k_n \cdot A^n \tag{5.18}$$

where k_0, k_1, k_2, ..., k_n are the coefficients. This is a generalization of the linear case discussed above. Thakur (1991) gives an example of a reaction that was well fitted with a cubic polynomial. Brown and Rothery (1993) also discuss polynomial models, describe in detail how to do polynomial curve fitting, and make three important points about polynomials: (1) they are very flexible over the fitted range and can be used as empirical models in a wide range of circumstances, (2) in general, it is not possible to ascribe a mechanistic meaning to the coefficients and (3) polynomial models must never be used for extrapolation outside the range of the fitted data, because these models may increase or decrease wildly outside the fitted range.

C. Logarithmic Dose–Response Curves

In the area of anticancer and antimicrobial chemotherapy, it is commonly assumed that the dose–response relationship is logarithmic (Skipper, 1986). With a logarithmic relationship, halving the drug dose will decrease the cell kill per unit time by a factor of $\sqrt{2}$. The logarithmic dose–response curve will occur when the drug-target cell interaction follows second-order kinetics:

$$\frac{dC}{dt} = -k \cdot C \cdot A \tag{5.19}$$

that is, the rate of cell killing is proportional to the dose (A) and the cell count (C), with k a second-order rate constant. As previously demonstrated (p. 292 in Jackson, 1992), if the drug has to give relatively few "hits" to cause cell death (say <100 per cell) then we may assume that cell killing does not give significant depletion of drug, so the kinetics will be apparent first order:

$$C_t = C_0 e^{(-kAt)} \tag{5.20}$$

where kA is the apparent first order rate constant, C_0 is initial cell count, and C_t is cell count at time t. Thus, cell kill will be given by:

$$f = C_0 [1 - e^{(-kAt)}] \qquad (5.21)$$

and $\log_{10}$ cell kill = $kAt/2.303$, so that log kill/dose = $kt/2.303$. Thus, a plot of log kill against dose would be linear with a slope of $kt/2.303$.

If the drug concentration is not constant, but decreases with time, which will usually be the case *in vivo*, then A must be replaced by its average value A',

$$A' = \frac{(A_0 - A_t)}{\ln(A_0/A_t)} \qquad (5.22)$$

where A_0 is drug concentration at zero time and A_t is drug concentration at time t. If we now redefine the rate of cell kill as

$$\frac{dC}{dt} = -k'C \qquad (5.23)$$

where k' = apparent rate constant = kA', it can be seen that the value of k' will depend upon the half-life of the drug, but the plot of log kill against dose will still be linear. Thus, the logarithmic dose–response curve is probably a reasonable description, both *in vitro* and *in vivo*, for those drugs that give pseudo-first order kinetics of cytotoxicity. It is believed that drugs that act by damaging DNA, for example, follow this pattern of kinetics (Skipper et al., 1970).

D. Log–Log Dose–Response Curves

With a log–linear dose–response relationship, as described above, if we halve the dose, the log kill is halved. Some drugs, particularly *in vitro*, seem more closely to approximate a log–log relationship. In this case, if we halve the dose, the log kill is reduced by log 2.

$$\log f = (\log A)/k \qquad (5.24)$$

where k is the IC_{90}, i.e., the dose that gives 1 log kill.

E. Multiexponential Dose–Response Curves

$$f = C_0 [1 - (F_1 e^{(-k1At)} + F_2 e^{(-k2At)} + \ldots + F_n e^{(-knAt)})] \qquad (5.25)$$

This is clearly related to Equation 5.21, but the single exponential term has been replaced by a complex expression with decay constants $k_1, k_2, \ldots, k_n$ and coefficients $F_1, F_2, \ldots, F_n$.

F. Hyperbolic Dose–Response Curves

The Hill equation (Equation 5.1), as noted above, is an expression of the law of mass action and is a convenient representation for effects of drugs that bind, reversibly and saturably, to sites on an enzyme or receptor. The Hill equation has three parameters: V_{max} is the intrinsic activity at complete saturation of the enzyme or receptor with its ligand (V_{max} reduces to 100 if results are expressed as percent of the unaffected value), K_m, or K_a, is the ligand concentration at half saturation, and n is the slope parameter. As noted above, when $n = 1$, a common experimental situation, the equation may be reduced to the Michaelis–Menten equation, or to its equivalent, the Scatchard equation (Equation 5.3). Systems that operate low on the saturation curve approximate a first order exponential (logarithmic) dose–response relationship. The hyperbolic model with $n = 1$ is often referred to in the pharmacodynamic literature as the E_{max} model, and when $n > 1$, as the sigmoid E_{max} model.

G. Logistic Dose–Response Curves

If the probability of a biological effect is binomially distributed, then it may be described by a logistic dose–response curve:

$$f = P(A) = \frac{\exp(a + bA)}{1 + \exp(a + bA)} \tag{5.26}$$

where, as in previous equations, f is the fraction affected and A is drug dose. From this it follows that:

$$\frac{f}{1 - f} = \exp(a + bA) \tag{5.27}$$

so that:

$$\ln\left[\frac{f}{1 - f}\right] = a + bA \tag{5.28}$$

This linearized version is known as the logit transformation. The logistic function follows a sigmoidal curve, with an inflection point at $x = -a/b$ (at which point $f = 0.5$). Mathematically, the logistic model is a variant of the Hill equation (Equation 5.5). Whereas the Hill equation is used to describe continuous variables (degree of enzyme inhibition and fraction of receptor occupied), Equation 5.26 describes binary events (whether or not an animal is asleep or whether or not a neurone fired). However, the logistic model is commonly used to fit radioimmunoassay (RIA) data, since RIA response curves are often sigmoidal. An example given by Thakur (1991) illustrates the four-parameter logistic model:

$$f = (a - d)/[1 + (A/c)^b] + d \tag{5.29}$$

where a = expected response at A = 0; b = slope of the logit plot; c = EC_{50}, the dose at which the response is (a + d)/2; and d = nonspecific binding.

H. Probit Dose–Response Curves

An excellent discussion of the use of probits is given in the classical work by Riggs (1963). The probit transformation is widely used to study quantal (all or none) events, though, as Riggs points out, it can also be used to describe graded responses. The probit dose–response relationship is used for events that follow a log-normal distribution. A very common example of its use in pharmacology and toxicology is to express lethality of drugs and toxins, since the the tolerance of mice (or other animals) to drugs or toxins is usually normally distributed with respect to the logarithm of the dose. Such dose–response curves are characterized by two parameters, the median effect point (described as the LD_{50} for the case of a lethality plot) and the slope of the probit plot, which is the reciprocal of the standard deviation (in log units). For modeling purposes we wish to predict lethality (or some other effect) from drug dose. Consider an example (adapted from Riggs, 1963) where the LD_{50} was 11.5 mg/kg and the standard deviation was 0.155 log. Suppose we wish to determine the percent lethality of a dose of 8 mg/kg. The procedure is as follows: (1) Take the log dose divided by the LD_{50}: log(8/11.5) = –0.1576; (2) divide the result of step 1 by the standard deviation: –0.1576/0.155 = –1.017; (3) using the result of step 2 as argument, read off the percent lethality from a table of areas under the normal curve = 0.155 or 15.5%. Thus, 8 mg/kg represents an $LD_{15.5}$ for this particular drug. Since owners of computers do not expect to look up numbers in tables, the area under the normal curve is calculated as follows: let the argument (the value calculated in step 2, –1.017 in the example) = X. Then

$$T = 1/(1 + 0.2316419 \cdot X)$$

$$P = \exp(-X^2/2)/\sqrt{(2\pi)}$$

$$f = 1 - P \cdot (F_1T + F_2T^2 + F_3T^3 + F_4T^4 + F_5T^5) \qquad (5.30)$$

where the coefficients are $F_1 = 0.31938153$, $F_2 = -0.356563782$, $F_3 = 1.781477937$, $F_4 = -1.821255978$, and $F_5 = 1.330274429$.

The area f is a fraction of one, so is multiplied by 100 to give percent. This polynomial approximation to the Gaussian probability function was taken from a publication from Hewlett Packard (1974) and gives excellent approximations for all positive arguments. For negative values of X, the argument is multiplied by –1, and the computed value of f is then subtracted from 1.

I. All-or-None Dose–Response Relationships

While the concept of a graded dose–response relationship is one of the cornerstones of pharmacodynamics, all-or-none responses do occur in biology — a notable example being membrane depolarization in nerve cells. The kinetics of inhibitors

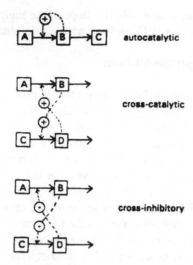

Figure 5.1 Reaction schemes that may exist in multiple steady states. In each case, A, B, C, and D are substrates. Solid lines indicate interconversion reactions, dashed lines with a plus sign show positive feedback effects, and dashed lines wih a minus sign show sites of negative feedback.

with all-or-none dose–response curves have been discussed (Jackson, 1992). It is characteristic of such inhibitors that they act at sites of positive feedback (or its kinetic equivalent, such as a cross-catalytic situation, as shown in Figure 5.1). Figure 5.2 shows a simple branched biochemical pathway that contains a positive feedback site and can be inhibited at various sites. In Figure 5.3, we see first the effect of a classical inhibitor acting on enzyme 2, then, in sharp contrast, the all-or-none dose response curve of a switch inhibitor that acts at the positive feedback site, and finally the effect of the switch inhibitor after disabling the positive feedback effect. The significance of positive feedback in biochemistry is that effectors that act at these sites (whether natural effectors or pharmacological agents) may act as on/off switches. Switch inhibitors differ from classical inhibitors in four respects:

1. As shown in Figure 5.3, classical inhibitors give a continuous dose–response curve, while switch inhibitors give a discontinuous, all-or-none dose–response curve.
2. Classical inhibitors require amounts that are at least stoichiometric (and often much more) with their target enzyme to switch off a pathway, whereas switch inhibitors can act at substoichiometric levels.
3. For classical inhibitors (if they do not bind covalently to their target) removal of the inhibitor reverses the inhibition, while systems inhibited by switch inhibitors display the property of hysteresis, and may remain switched off after the inhibitor is removed, i.e., they act as biochemical memories.
4. Classical inhibitors may perturb their target pathway to a new, inhibited steady state, or away from any possible steady state. In the former case, the resulting steady state need not be a physiological one (i.e., a state in which the organism normally exists), and in the latter case it certainly is not. In contrast, the effect of switch inhibitors is to flip the system from one physiological state to another: induced vs. uninduced, proliferating vs. nonproliferating, cycling vs. quiescent (Jackson, 1992).

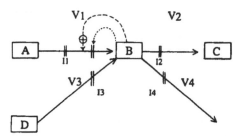

Figure 5.2 A branched linear pathway in which the primary source reaction is subject to feedback activation by low concentrations of its product, B, and to feedback inhibition by high concentrations of B. The dashed line indicates activation and the dotted line indicates inhibition.

J. The Bell-Shaped Dose–Response Curve

A number of ligands show biphasic dose–response curves, a phenomenon analogous to excess substrate inhibition in enzyme kinetics. In that case, the substrate saturation curve passes through a maximum and then declines. In the case of an inhibitor, the activity curve reaches a minimum and then starts to increase. The usual mechanistic explanation is that the inhibitor is acting at two sites. For example, some drugs that kill cells in early S phase, at higher concentrations may cause a nonlethal block of progression from G_1 phase into S phase and thus antagonize their own cytotoxicity. One rate equation that shows a biphasic curve is

$$f = [A/P_1/(1 + A/P_1)]/(1 + (A/P_2)^2) \tag{5.31}$$

where P_1 and P_2 are the binding parameters; e.g., if $P_1 = 1.5$ and $P_2 = 200$, then the curve reaches a minimum at $A = 31$ ($f = 0.931$) and at $A = 600$, $f < 0.1$. Note that a second-order polynomial (Equation 5.18) can also give biphasic curves, but in the latter case f is not asymptotic to 0.

VI. TIME-DEPENDENCE OF PHARMACODYNAMIC EFFECTS

For many drugs, effect is a function of both concentration and time, and it is sometimes stated that effect is proportional to $C \times t$ (where C is concentration and t is time). Millenbaugh et al. (1995) noted that a more generally applicable relationship is given by:

$$s = C^n \cdot t \tag{5.32}$$

where s is a constant denoting drug sensitivity for a particular system and n is a pharmacodynamic exponent that relates the relative importance of concentration and time in determining drug effects. When $n > 1$, concentration is the primary determinant of the drug effect, and when $n < 1$ the effect is primarily time dependent. Millenbaugh et al. suggest that the value of the n parameter has implications for

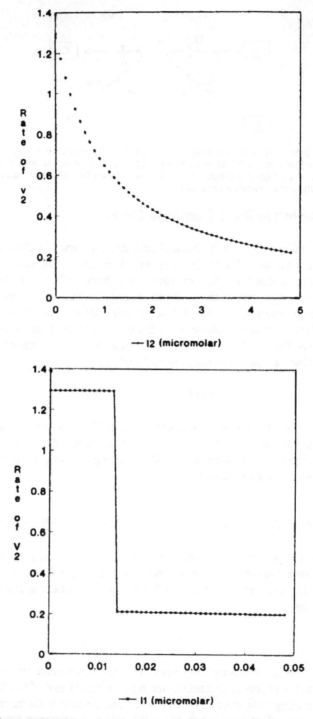

Figure 5.3 Dose–response curves for effects of inhibitors on flux through the pathway of Figure 5.2. (A) Effect of inhibitor 2. (B) Effect of inhibitor 1 on the intact system. (C) Effect of inhibitor 1 on the system when positive feedback is disabled.

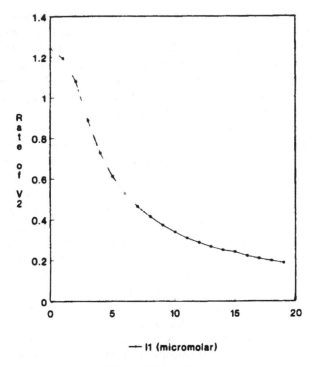

— I1 (micromolar)

Figure 5.3 (continued)

selection of optimal clinical dosage regimens. For instance, when $n > 1$, bolus administration should be more effective than an infusion. Equation 5.32 may be fitted to experimental data by either a two-step process, in which IC_{50} values are calculated for each exposure time and the IC_{50} estimates (representing C) are fitted to Equation 5.32, or, alternatively, by a nonlinear regression fit to

$$f = C^m/[(s/t) \wedge (m/n) + C^m] \tag{5.33}$$

where f is the fraction affected, m is the Hill coefficient, and C, t, n, and s are as defined for Equation 3.31.

VII. PROGRAMS FOR SIMULATION OF BIOCHEMICAL PATHWAYS

If our pharmacodynamic model consists of a single equation describing a dose–response relationship, then one of the equations described above will probably give an adequate description. However, for some kinds of problems it is necessary to describe a whole biochemical pathway; for example, if we are studying interactions of two drugs that block the same pathway at different sites, the only practical way to predict the drug–drug interactions is by modeling the target pathway. Although the science of enzyme kinetics is over 80 years old and is described in many excellent textbooks, the kinetics of multienzyme systems has been much less thoroughly explored. With the exception of very simple systems of two or three enzymes, multienzyme systems display complex behavior that is hard to describe in terms of explicit rate equations. Because of the enormous variety of possible system configurations, it is hard to draw useful generalizations. However, with the help of a computer it is possible to create a kinetic model of any required system, i.e., a set of rate equations describing the kinetics of that particular system. Analysis of the model can predict rates of individual reactions within the system. Then the set of equations can be integrated to determine concentrations of substrates and products at any required time. Effects of inhibitors and activators can be predicted, and the regulatory properties of the system can be dissected out. Armchair biochemistry experiments can be performed to answer such questions as, what happens to the control of the system if this particular feedback effect is removed?

Although our primary concern is with kinetics, certain important thermodynamic concepts need to be emphasized. First, life is a property of systems that are operating far from chemical equilibrium and can only be described in terms of nonequilibrium thermodynamics. Living systems are open systems, whose existence is dependent upon exchange of matter with the environment, and whose distance from equilibrium must be sustained by continual dissipation of energy. Second, living systems spend much of their time at or close to a steady state. While these steady states are not chemical equilibria (at which entropy production is zero), they share some characteristics with equilibrium systems, such as the constancy of their chemical composition with time, and the fact that entropy production at these steady states is at a local minimum. Third, another essential characteristic of life is that the system does not remain indefinitely in a single steady state, but can be switched (in response to a chemical or physical stimulus) from one steady state to another. This final thermodynamic aspect of living organisms is a reflection of their **organized complexity**. Here no attempt is made to provide a rigorous definition of complexity as the word is used by systems theorists, but note that the term implies systems containing a large number of variables and multiple nonlinear relationships between those variables.

A model of a biochemical pathway consists of terms that represent reactant concentrations (which may be constant or variable) and rate expressions for the reactions that interconvert the reactants. The rate expressions generally describe enzyme-catalyzed reactions, though they may be nonenzymatic reactions or transport processes. Thus, the reactant concentrations at any required time may be computed by integrating the system of rate equations. The use of computers to model biochemical pathways goes back to the early days of computing. Among the pioneers in this field were Britton Chance and his associates at the University of Pennsylvania, which was one of the birthplaces of the first-generation computers. Notable among these early contributions was the model of Garfinkel and Hess (1964) of glycolysis in red blood cells.

Perhaps the primary application of metabolic modeling has been to study the regulation of biochemical pathways. Control points and rate-determining steps can be identified, the significance of negative and positive feedback effects can be explored, and responses to addition of various metabolic intermediates can be predicted.

Modeling provides a method (sometimes the only method) of testing hypotheses; e.g., if activity of a particular enzyme is abnormally low in a mutant cell, how should this affect flux through the pathway? Many of the existing biochemical models have specifically addressed the prediction of inhibitor effects: what is the most sensitive site for inhibition in a particular pathway? How much must that site be inhibited to decrease total pathway output by (say) 90%? What will the shape of the dose–response curve be? How do the substrate pools respond to the presence of inhibitors?

Quite a large number of special-purpose biochemical models have been created to describe particular biochemical pathways in various organisms. Kinetic simulation has been used to model glycolysis (Schellenberger et al., 1981), the pentose shunt (Thorburn and Kuchel, 1985), tryptophan metabolism (Heinmets, 1971), the urea cycle (Bachmann and Colombo, 1981), pyrimidine metabolism (Jackson, 1986), purine metabolism (Bartel and Holzhötter, 1990), and cyclic AMP signaling (Segel, 1984). Seven different authors or groups have modeled folate cofactor reactions and responses of the folate pathways to antifolate drugs, with a generally excellent degree of consistency in their conclusions (Werkheiser et al., 1973; Jackson and Harrap, 1973; Harvey and Dev, 1975; Vorontsov et al., 1980; White, 1986; Seither et al., 1989; Morrison and Allegra, 1989). Several authors have modeled various aspects of deoxyribonucleotide metabolism and the control of the deoxyribonucleoside triphosphate (dNTP) pool composition, which is regulated by a remarkably complex system of positive and negative feedback effects (Grindey et al., 1975; Morrison et al., 1975; Nicolini et al., 1977; Jackson, 1989). Most of these papers compare their predictions with experimental data.

A brief description of two of these models may help to illustrate some of the things we may learn from modeling studies. Bartel and Holzhütter's (1990) model of purine in the rat liver includes 26 substrates (purine bases, nucleosides, nucleotides, oxidized and reduced pyridine nucleotides, sugar phosphates, and magnesium ion), 24 rate processes (mainly enzyme-catalyzed reactions and some uptake and efflux processes), and four rapid equilibria. This model is thus well within the capacity of BioCHIMICA (described below) to emulate, and the authors provide enough information to enable readers to repeat their calculations. Using their model, Bartel and Holzhütter simulated the effect of hypoxia (by setting ATP production to zero) and showed that the predicted changes in pools of ATP, ADP, AMP, and IMP were in good agreement with experiment. The predicted changes in free adenosine concentration, however, did not agree well with experiment, since the model predicted a rise in free adenosine for about 40 min after the onset of hypoxia, while experiment showed a peak at 5 to 10 min. This suggested either that the assumed inhibition constant for inhibition of 5'-nucleotidase by ATP was too low or that some other regulatory process, currently unknown and thus not included in the model, is operating. When Bartel and Holzhütter modeled reoxygenation, the predicted nucleotide and nucleoside pools moved back in the direction of their normal values. However, if the duration of hypoxia was greater than 30 min, the system was unable to recover completely. This study of rat liver purine metabolism is interesting for three reasons. First, many of the predicted properties of the model system were in close agreement with experiment, showing that the general structure of the model was probably reasonably valid. Second, the fact that the model did not accurately predict the changes in adenosine pools in hypoxia is an example of how mathematical models can serve a useful purpose by pointing to errors or gaps in our experimental data. Third, the inability of the simulated system to recover completely from long periods of hypoxia, and thus, in a sense, to predict death, shows that this model is sufficiently complex to show interesting behavior.

Another model, over 20 years old, but still an excellent example of the use of simulation to address complex questions about complex systems, is the study by Barbara E. Wright (1973) of differentiation in the slime mould *Dictyostelium discoideum*. This study of reactions of carbohydrate metabolism used a program called METASIM that could simulate up to 25 reactants and 25 enzymes, written in PL1 for the IBM-360. The model predicted that certain enzymes were able to vary within wide limits without affecting the process of differentiation (e.g., UDPG pyrophosphorylase), while six other variables (e.g., trehalose synthesis) were critical to the differentiation process. Wright also stressed the importance of key *in vivo* flux values in relation to cellular differentiation. One of her major conclusions was: "a system of coupled interacting biological components has properties beyond those apparent from the analysis of these components in isolation." There is nothing mystical or vitalistic about this concept of emergent properties. It simply says that for certain qualitative behaviors to be demonstrated requires a certain level of system complexity. If we want to reduce the effort required to lift a heavy weight with a piece of rope, it takes three pulley wheels; if we want an enzyme system to act as an on/off switch, there must be at least two feedback effects (one positive and one negative feedback; alternatively a switch can be constructed with three negative feedback effects). The general message to be learned from Wright's work is that experimental studies of biological systems often show a bewildering variety of biochemical changes during a process such as differentiation, and computer modeling is an excellent tool for dissecting out which of these changes are essential and which are incidental.

One of the author's early models (Jackson, 1980) described the folate cofactor pathways as a system of 63 reactions, involving 37 variables. Among other things, this model was used to analyze biochemical changes that may contribute to resistance of tumor cells to the antifolate drug, methotrexate. Seven different factors that contribute to methotrexate resistance were considered. Here again a computer model provided a means of analyzing complexity and address questions that were hard to approach in any other way.

Although kinetic modeling has now been applied to an impressive array of biochemical problems and has repeatedly proved its value as a tool for the analysis of biochemical complexity, there remain many problems, potentially highly amenable to this approach, which have yet to be modeled. Resistance to antiviral drugs (particularly antimetabolites) presents many analogies to anticancer drug resistance. Modeling the process of resistance of, for example, the AIDS virus to azidothymidine could assist in the design of protocols to delay the onset of resistance, of design of drug combinations that might maximize killing of resistant viruses, or the use of biochemical modulators to overcome drug resistance.

Useful models have been constructed of several areas of metabolism, but other equally important pathways are essentially untouched. To what extent is protein synthesis affected by amino acid availability? While we know a lot about regulation of the pathways of amino acid synthesis, matching the cell's needs for 20 different protein precursors must require a remarkable balancing act, and we have only a rather vague idea how this is done. This is certainly a level of complexity that will require a good model to make much sense of it. Another area of cellular biochemistry that has not been much modeled, and would certainly benefit from the approach, is the signal transduction pathways. These complex, overlapping, multiple systems of protein phosphorylation and dephosphorylation clearly act as cellular switches, amplifiers, and volume controls. While the qualitative reactions involved in these pathways are being rapidly elucidated, the *kinetics* of these switches remain almost totally obscure. Given that signal transduction pathways are likely to comprise an important group of targets for future drug design, being able to estimate the effects of inhibiting these pathways at various points will surely be of pharmaceutical relevance. The signal transduction pathways generally terminate in proteins that bind to DNA and regulate transcription, the transcription factors. It has been estimated that mammalian cells may contain from 2000 to 10,000 distinct transcription factors, and faced with the task of keeping these straight in their heads, even the most skeptical are likely to concede the usefulness of computer modeling.

A. The BioCHIMICA Model

Most of the biochemical modeling studies published to date have used special-purpose models written specially for the particular pathway under study. These models are obviously tailored to the needs of their particular pathway, and (other things being equal) they are likely to run faster than a general-purpose model. Their disadvantage, obviously, is that making any changes to such models generally requires reprogramming, and this requires familiarity with the code. BioCHIMICA, by contrast, is a general purpose biochemical modeling program that can be used to model any linked reaction system and requires no programming by the user (Jackson, 1995). It consists of four modules, which interact as summarized in Figure 5.4. A pathway description module, PATHEDIT (module 1 in Figure 5.1), asks the user which reactions are to be described in the model and how they interconnect. In the present implementation, up to 100 reactions (enzymatic reactions, nonenzymatic chemical reactions, or diffusion processes), involving up to 96 substrates and up to 64 inhibitors, may be modeled. Each reaction may be described by any rate equation from the rate equation library, which currently contains 24 rate equations and allows use of up to eight user-defined equations. In addition to being connected by rate processes, two or more substrates may be treated as being in instantaneous equilibrium. Enzyme levels may be treated as variable, and substrates may also be inhibitors. The rate processes take place in, or between, up to ten spatial compartments.

When an enzyme name has been entered, the PATHEDIT module checks the knowledge base (module 4 in Figure 5.4) to see if this enzyme (in the appropriate species and tissue) is known to it. If it is, the existing data may be used unchanged,

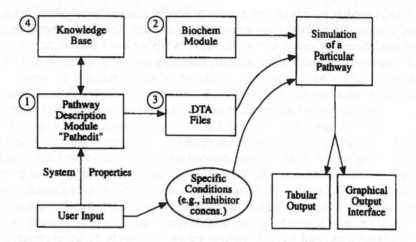

Figure 5.4 Block diagram of the BioCHIMICA biochemical modeling software package.

or edited as necessary; if not, the program asks the user which rate equation is to be used and for the appropriate parameter values (up to 14 for each enzyme). Based on a combination of user input and information from the knowledge base, the PATHEDIT module creates a descriptor file (module 3 in Figure 5.4) for the particular pathway to be modeled. This file is passed to the differential equation solver, BIOCHEM. This module (number 2, in Figure 5.4) allows the user to specify the time period to be simulated and the frequency of output. The BIOCHEM module then solves the rate equations for the particular set of parameter values contained in the descriptor file.

In summary, most biochemical modeling to date has been done using special-purpose programs written for each particular application. Writing these programs is time-consuming, and requires biomathematical and programming expertise. Bio-CHIMICA, by contrast, requires no biomathematical or programming skills, and will run on any IBM-compatible personal computer. It is menu-driven and user-friendly and makes biochemical modeling accessible to any biochemist.

VIII. MODELS THAT COMBINE PHARMACOKINETICS AND PHARMACODYNAMICS

We have considered ten empirical equations relating drug effect to drug concentration, as well as detailed biochemical models that may include entire pathways. On their own, these equations or systems of equations do not, however, take into account the fact that the concentration of drug is not a constant, but may be undergoing constant change. Of course there are artificial situations in the laboratory where drug concentration may be considered constant, such as bacteria or mammalian tissue culture cells growing *in vitro* or an isolated organ being perfused with a constant drug concentration. However, to study the *in vivo* situation we need to consider models that combine expressions for pharmacokinetics and pharmacodynamics, which are referred to as PK/PD models. A model that describes drug effects as a function of time, but which does not necessarily model PK explicity, is termed a kinetic effect model, or KEM (Grevel, 1987). We will treat the terms KEM and PK/PD as synonymous.

In the simplest form of PK/PD model, the drug is assumed to act in one of the compartments of a classical PK model. For example, if we assume that an antibiotic is acting primarily to destroy bacteria in the blood, then it is only necessary to couple the appropriate dose–response relationship to a one-compartment PK model (or to the central compartment of a multiexponential model). This approach was used by Coltart and Shand (1970) in an early PK/PD model of action of beta-blockers. Wagner et al.(1968) described a model where the effect of lysergic acid diethylamide (LSD) on arithmetic test scores was related to the concentration of drug in the peripheral compartment of a classical two-compartment open PK model.

For many drugs, the peak effect is delayed in time after the peak plasma level; also, drug effects may often be measured when drug can no longer be measured in plasma. This behavior may be modeled by assuming that the drug acts in a hypothetical effect compartment, in which drug is assumed to enter and leave by first order kinetics. The volume of the effect compartment is assumed to be sufficiently small that it does not perturb the kinetics of the other compartments. This approach was pioneered by Sheiner et al. (1979); see also Holford and Sheiner (1981). It has been used to generate PK/PD models of a wide variety of drugs, including cholinesterase inhibitors (e.g., d-tubocurarine), alpha-agonists (e.g., clonidine), benzodiazepines (e.g., midazolam), and dopamine agonists (e.g., bromcriptine). Brief reviews of this approach have been given by Grevel (1987) and Wagner (1993). A number of PK/PD models of anticancer drug action are reviewed by Kobayashi et al. (1993). Readers who wish to construct such a model may use the MKMODEL program (Holford, 1990; see Appendix) for this purpose.

IX. COMPUTER PROGRAMS FOR COMBINED PK/PD MODELING

A number of published studies have utilized the ADAPT II program of D'Arginio and Schumitsky (1979) for parameter estimation and simulation in PK/PD systems. Kobayashi et al., (1993) give seven published examples of the use of ADAPT II for analysis of anticancer drug studies. These authors also discuss the use of NONMEM (the nonlinear mixed effect model of Beal and Sheiner, 1989). This program is particularly suitable for obtaining estimates of population variability in clinical studies where data sets may be sparse and may have missing points. NONMEM is said to require a considerable degree of mathematical skill and familiarity with advanced statistics (Kobayashi et al., 1993), though the same authors cite a number of examples where it has been applied to clinical pharmacology studies.

PCNONLIN (Metzler and Weiner, 1992) is a small, PC-based package intended primarily for nonlinear model fitting, particularly of pharmacokinetic data. When data have been fitted to a classical PK model, PCNONLIN is very useful for predicting blood levels after differing doses and schedules of administration. However, PCNONLIN can fit effect data to any of a library of eight pharmacodynamic models or four very simple PK/PD models (all of which have Michaelis–Menten output). It will also accommodate user-defined models that may consist of up to ten differential equations.

MLAB is another nonlinear model-fitting package, in this case a rather powerful general-purpose system. The MLAB Applications Manual does not give examples of PK/PD models, but it gives two examples of pharmacological (i.e., PK) models, one of which includes a delay term. MLAB clearly is entirely suitable for constructing simple or moderately complex PK/PD models.

Another commercially available package is MKMODEL. Unlike some of the packages discussed above, MKMODEL is specifically written for fitting pharmacological data. The package includes a library of six common PK models, and user-defined models can also be created. Simple pharmacodynamic modeling is also possible with MKMODEL, and user-defined PD models can be constructed. MKMODEL will model effect-compartment PD systems, and the manual gives an example of a PK/PD model of frusemide. Any of the six standard PK models may be utilized in the effect-compartment model. The PD function is restricted to a sigmoid E_{max} equation.

The P-Pharm program (see Appendix) is a software package for modeling PK and PD data with an emphasis on population modeling. It is intended for use with a relational data base, such as ORACLE, and can select data for specified subpopulations. It calculates mean population parameter values, together with their variance. For each individual, one or more drug concentration samples must be supplied, together with a set of measurements of covariates, such as age, weight, or creatinine clearance. P-Pharm will search for an optimal linear relationship between the model parameters and the covariates. P-Pharm contains an option for computing individualized initial dosing regimens for new patients by applying past experience using the population pharmacokinetic model. The model calculates initial estimates and can then factor in user-defined therapeutic constraints to evaluate the optimal dosage regimen.

X. MODELS OF INDIRECT PHARMACODYNAMIC RESPONSES

It has been argued by Jusko and his colleagues (Dayneka et al., 1993) that when drugs act through a slow accretion of a response that results indirectly from the cumulative effect of a drug on inhibition or stimulation of some factor, the pharmacodynamics may differ markedly from direct drug effects. They suggest that although use of an effect compartment may sometimes allow a reasonably good description of pharmacodynamics over a limited concentration range, such an approach is an approximation at best. So far the PD models that have been discussed in the present chapter have mostly been direct models, though the detailed simulations of drug effects on biochemical pathways modeled by programs such as BioChimica can accurately describe indirect effects. Dayneka et al. (1993) compare four basic indirect models, and discuss their likely applications.

XI. MORE COMPLEX EXAMPLES OF
PHARMACOKINETIC/PHARMACODYNAMIC MODELS

The relatively simple programs discussed above are all primarily model-fitting packages that are mostly intended for fitting classical PK models to measured levels of drug in blood, though, as noted, some of these packages can also fit simple PD models. When we attempt to construct more complex PK/PD models, e.g., a model consisting of a PB-PK model coupled to a simulation of a biochemical pathway, it becomes essential to separate the function of parameter estimation from the function of simulation. These models may contain hundreds of parameters, and it would be totally impractical to attempt to estimate them simultaneously from a single data set, nor does it make sense to attempt this, seeing that many of the values are known from other sources. Gallo (1991) has discussed the subject of parameter estimation for PB-PK models; this article discusses how data from the literature, from *in vitro* studies, and from *in vivo* measurements can be utilized, often in the same model. The subject of parameter estimation for PB-PK models is covered in Chapter 2. Briefly, it is usually convenient to use literature values for total tissue volume, organ blood flow, and for fractions of a tissue that are extracellular and intracellular. *In vitro* methods are often used to measure tissue–plasma partition coefficients for drugs, though this information may be obtained from the *in vivo* ADME study. The method used to fit the PD parameters obviously depends on the nature of the PD end point: if it is a physiological or behavioral measurement then an *in vivo* experiment is necessary, and it may be the same study in which drug levels were measured.

For the system of a kinetic model of a biochemical pathway coupled to a PB-PK model, the PD parameters consist of enzyme kinetic data (V_{max}, K_m, and K_i values) plus steady–state pool measurements of biochemical intermediates. The kinetic parameters are almost always measured in completely or partially purified enzyme preparation; we discussed above the circumstances in which inhibition kinetics in the cell may differ from what is observed with isolated enzymes. Steady-state levels of substrates may be available from the literature or may have to be measured using appropriate biochemical assays.

Simulation of anticancer, antiviral, or antibacterial PD is often done at the whole-cell level: the main end point is cell kill, and we may not be interested in the biochemical details of what the drug is doing. In this case, our PD model can consist of an *in vitro* dose–response curve, such as one of the ten empirical dose–response curves described above. Combinations of both classical PK models and PB-PK models with both empirical PD models and with biochemical pathway kinetic models have been studied using a published program of the author (Jackson, 1992). The major modules of the EXCHEMO model of experimental chemotherapy are shown in Figure 5.5. The first module is the PK model and the second (inhibition kinetics) translates drug concentrations into cytokinetic effects. The third, cytokinetic, module is discussed in the next chapter.

The EXCHEMO package contains a model of 30 reactions of pyrimidine biochemistry (Figure 5.6); the model is called HEPATOMA because its primary function was to simulate pyrimidine biochemistry in liver cancer, but with appropriate parameter changes it can model these reactions in other tissues. HEPATOMA was used to

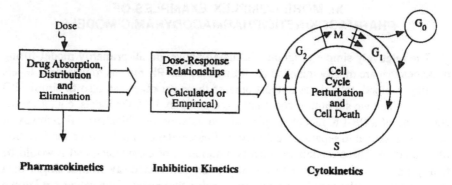

Figure 5.5 Block diagram of the VIROCHEM program for modeling pharmacokinetics and pharmacodynamics of antiviral drugs.

predict the effect of 5-fluorouracil (5-FU) on DNA and RNA synthesis in hepatoma cells. Several 5-FU concentrations were simulated, and the output was fed to the PD module of EXCHEMO (a program called DRCURVE). DRCURVE then constructed a dose–response curve, took the predicted 5-FU concentrations from the PK simulation, determined the inhibition of DNA and RNA biosynthesis (which result in S-phase cytotoxicity and G_1 arrest, respectively) and passed the results on to the cytokinetic module, which predicted the antitumor activity of the single dose of 50 mg/kg of 5-FU that had been modeled. The second stage of the simulation now factored in an additional biochemical observation: bone marrow is known to contain about 10 μM thymidine, which partially blocks the DNA-directed effect of 5-FU, though not the RNA-directed effect. Consequently, another series of HEPATOMA simulations were run, in which 10 μM thymidine was assumed to be present. Now a cytokinetic simulation was run, in which tumor was assumed to be exposed to 5-FU in the absence of thymidine, and bone marrow was exposed to 5-FU in the presence of thymidine. In the simulations described by Jackson (1992), the 5-FU concentrations were calculated by a three-compartment classical PK model, but the same approach can be used using tissue drug concentrations predicted by a PB-PK model, as our final example shows.

A PK/PD simulation that absolutely required detailed biochemical input was also discussed by Jackson (1992). The antipyrimidine N-phosphoacetyl-L-aspartate (PALA) inhibits an early stage of pyrimidine biosynthesis and depletes cellular pools of pyrimidine nucleotides. This, in turn, has the effect of increasing the rate of intracellular activation of 5-FU to its active metabolites, 5-FdUMP and 5-FUTP, which means that, in tissue culture, PALA is synergistic with 5-FU. The important question is, does this biochemical modulation improve the antitumor selectivity of 5-FU, or just make it more dose potent? Since a relatively small number of active drugs can be combined in an enormous number of ways, it is only possible to test a very small fraction of the possible combinations *in vivo*. Modeling is useful in this context as a way of suggesting which of the possible combinations would be of most interest to select for experimental study. In the EXCHEMO simulation, a three-compartment PK model was used to predict levels of both drugs in a mouse simultaneously treated with 5 mg/kg of 5-FU and 20 mg/kg of PALA. The combined

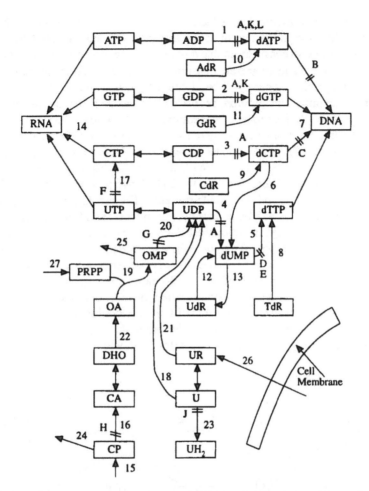

Figure 5.6   A detailed kinetic model of reactions of pyrimidine metabolism. The model includes 26 enzyme-catalyzed reactions (rates 1 to 25 and 27) and one active transport process (rate 26). The inhibitors modeled are A, hydroxyurea; B, ara-ATP; C, ara-CTP; D, 5-FdUMP; E, 5,8-didaza-10-propargylfolate; F, 3-deazauridine 5′-triphosphate; G, 6-azauridine 5′-phosphate; H, N-phosphonacetyl-L-aspartate; J, 5-diazouracil; K, 2-chloro-2′-deoxyadenosine; and L, ara-GTP. CA, carbamoylaspartate; CP, carbamoylphosphate; DHO, dihydroorotate; OA, orotic acid; U, uracil; UH₂, dihydrouracil. (From Jackson, 1986. Reproduced with permission of Pergamon Press.)

biochemical effect of the two agents was then simulated for a large number of time points, and the calculated cytokinetic perturbations used to determine the antitumor response.

A final example from the author's earlier book (Jackson, 1992) is worth discussing here because it provides an example that could only be modeled by a combination of a PB-PK model with biochemical pathway simulation. It addressed the question, what is the probable effect on tumor response to methotrexate if two different kinds of resistant mutant are seen, one in which there is a fivefold increase in activity of the target enzyme dihydrofolate reductase (DHFR) and the other in which there is a threefold decrease in the rate of active transport of methotrexate into the tumor cells? The model also considers double mutants, i.e., cells with both a fivefold DHFR overproduction and a methotrexate transport defect. As a final complication, we ask how the therapeutic response will be affected by metastatic tumor cells appearing in the brain, an important question, because methotrexate penetrates the blood–brain barrier poorly. A PB-PK model of methotrexate distribution in the mouse was used to calculate drug levels after a 10-mg/kg i.v. bolus dose. A biochemical model called FOLATE (Figure 5.7), which includes 22 reactions of folate cofactor interconversion and antifolate drug action, was used to calculate dose–response curves for four cell populations: the original sensitive cells, the DHFR-overproducing mutants, the transport-defective mutants, and the double mutants. The cytokinetic module then predicted drug effects in tumor, bone marrow, gut mucosa, and brain, using drug levels calculated for each compartment by the PB-PK model. The results were interesting: whereas the number of primary tumor cells was decreasing by 24 h after treatment, the number of metastatic cells started to increase after a very transient decline, because of the poor penetration of the blood–brain barrier by methotrexate. Even in the primary tumor, at 18 h after treatment when the numbers of drug-sensitive tumor cells and of both the resistant single mutants were declining, the number of double-resistant cells was already increasing. In the brain, the single-mutant resistant populations were already increasing by 16 h after treatment.

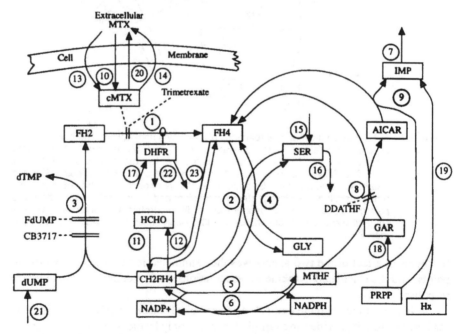

Figure 5.7 Diagram of reactions simulated by the FOLATE model. Abbreviations are CB3717, 5,8-dideaza-10-propargylfolate; CH$_2$FH$_4$, 5,10-methylenetetrahydrofolate; FH$_2$, dihydrofolate; FH$_4$, tetrahydrofolate; and MTHF, 10-formyltetrahydrofolate. The identities of the numbered reactions are listed in Jackson (1980).

This final example demonstrates the range of questions that can be simulated using the combination of a PB-PK model with a biochemical model of the target pathway of the drug under study. While it used to be necessary to devote dozens of hours of programming time to creating such models, the recent availability of general-purpose biochemical modeling packages such as BioCHIMICA has made this approach much more feasible.

With a modest investment in increased data collection and a more computer-intensive analysis of the resulting data, PK/PD models could provide valuable insights into the properties of developmental drugs. As once stated by a member of the U.S. FDA:

> At the present time, PK/PD techniques are used to establish required bioavailability and bioequivalence information. Beyond that, these techniques have seldom been used in such a way as to contribute to the efficiency of drug development. This is somewhat surprising since key PK/PD concepts and techniques have been evolving throughout the last 30 years.... If the efficiency of drug development can be improved through use of PK/PD, there are potential gains for all interested parties: patients, drug developers, and regulatory scientists. (Peck, 1993)

CHAPTER 6

Cytokinetic Modeling

I. PROLIFERATION KINETICS OF BACTERIA

When measuring the effects of chemotherapeutic agents (antibacterial, antiviral, antifungal, antiprotozoal, and anticancer drugs) the primary end point is killing cells of the pathogen. Many important aspects of drug action can be understood at the whole-cell level — dose–response relationships, selectivity, resistance, and cell cycle specificity. Cell proliferation, cell killing, and cytostasis can all be measured *in vitro*, and these *in vitro* parameter values may then be used to predict the course of the disease and its treatment *in vivo*. The science of cytokinetics provides quantitative tools for the study of these questions. The simplest kinds of cytokinetic model assume a homogeneous pathogen population, with a uniform cell doubling time (and thus exponential growth). Even an unrealistically simple model, neglecting effects of the immune system, acquired drug resistance, heterogeneity of doubling times, and noncycling cell populations, can be useful for asking some kinds of questions. For example, if we have an antimicrobial drug that can eliminate 3 logs of pathogen at its maximum tolerated daily dose, and survivors grow back with a 2.5-h doubling time, and if 11 logs of pathogen cells constitute a lethal dose, what is the largest initial inoculum that can be completely eliminated by daily treatments, and how many treatments will be required? These questions may be studied using the ANTI-BACT program (see Appendix). This is a program that simulates effects of antibacterial drugs, administered singly or in combination, at doses and times chosen by the user, on exponentially growing populations of bacteria. Sample output from ANTIBACT is shown in Figure 6.1. Starting with an inoculum of 100 cells, a regimen that gave three log kill every 24 h (beginning at 24 h) eliminated the infection in 19 d. If the inoculum was increased to 2000 cells, the infection could be eliminated in 30 treatments. Larger inocula could not be controlled by 30 daily treatments. If the doubling time was increased to 3 h, even very large inocula (anything that did not cause death before the first treatment at 24 h) could be controlled by not more than 15 daily treatments. Conversely, if the doubling time was decreased to 2 h,

127

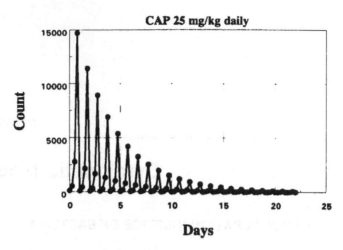

Figure 6.1 Simulation with the ANTIBACT program of treatment of a bacterial infection with 20 daily doses of chloramphenicol.

treatment with a regimen that gave three log kill every 24 h could not control bacterial proliferation, and even an inoculum of one cell was lethal.

These calculations may be criticized on the grounds that the assumptions are unrealistically simplistic, and also because the same results may readily be reached without a computer, using a pocket calculator. In fact, the ANTIBACT model may readily accommodate much more complex assumptions. The calculations discussed above were made after turning the mutation rate for development of drug resistance to zero and disabling the simulation of the immune response. If we factor in destruction of bacteria by the immune system and acquired drug resistance, the system becomes too complex to solve manually.

The use of computer models for prediction of drug resistance is discussed in Chapter 7. However, modeling of the immune response is an appropriate topic for our present concern with cytokinetics. The ANTIBACT model contains a simple model of the immune response, which assumes that a stimulated lymphocyte population proliferates with a rate that depends upon (1) the existing number of stimulated lymphocytes, (2) the size of the antigen challenge (e.g., the number of bacterial cells), and (3) the degree of antigenicity of the pathogen. Killing of bacteria is then treated as a function of time and of the size of the population of stimulated lymphocytes. Although this is strictly a model of cellular immunity, it seems likely that humoral immunity will follow similar kinetics. Consider a simulation with a moderately immunogenic organism (immunogenicity factor set to 36%) with a doubling time of 3 h. Following infection with an inoculum of 100 cells, the organism starts to replicate and stimulates replication of a sensitized lymphocyte clone. By 65 h the bacterial count begins to decline, and the infection is eliminated by 76 h. Even a weak immune response can have a major influence on the outcome of antibacterial treatment. As discussed above, if the pathogen doubling time is 2 h, a drug that gives three log kill, administered daily, is unable to control bacterial proliferation. However, if immunogenicity is set to 11% the results are surprisingly different. This weak immune response has a negligible influence on bacterial growth in the absence of drugs — an inoculum of two cells can be eliminated, but an inoculum of three cells is ultimately lethal. When this system is treated with a daily dose of antibiotic that gives three log kill, an inoculum of 1000 cells was eliminated in 53.75 h.

Since real-life drugs may be toxic, there is a limit to the dose of drug that may be administered or to the number of treatments that may be given. ANTIBACT models three kinds of toxicity. First, myelosuppression (bone marrow toxicity) is simulated by modeling drug-induced kill of bone marrow stem cells and allowing the surviving bone marrow population to recover with a doubling time that may be set to any required value. Typically, this would be about 18 h for a mouse and 36 h for a human. If the net kill of bone marrow stem cells exceeds a preset value (usually two logs) the animal or patient is assumed to have succumbed to lethal myelosuppression and the simulation stops. Second, immunosuppression is simulated by modeling drug-induced kill of immune lymphocytes; again these may recover with a doubling time that may be set to any value appropriate for the species being modeled. Here too, there is an adjustable parameter that determines the tolerable limit of immunosuppression. The third kind of toxicity that is modeled by ANTI-BACT is cumulative toxicity. For some drugs (aspirin and penicillin-G) there is no evidence of cumulative toxicity and no known limit to the lifetime dose that can be tolerated. Other drugs (e.g., Adriamycin) appear to cause cumulative toxicity so that the total lifetime dose must be kept within the tolerable limit. In the case of Adriamycin, the cumulative toxicity is to the heart, but other drugs may cause cumulative renal toxicity or neurotoxicity. ANTIBACT does not distinguish between the different sites of cumulative damage, but it does keep track of the cumulative dose and prints a warning if it would exceed the tolerable limit for that particular drug.

The assumption that bacterial killing follows a log–linear dose–response relationship is a reasonable one for many drugs. However, the ANTIBACT program is capable of modeling any of the top ten empirical dose–response curves described in Chapter 5. In fact, for prediction of multilog bacterial cell kill, the Hill equation, in the following form, is probably the most widely used model:

$$f = 1 - \{1/(1 + [C/IC_{50}]^n)\} \tag{6.1}$$

where f, in this instance, is the fraction of the bacterial population killed by a drug concentration C, IC_{50} is the mean lethal dose, and n, the Hill coefficient, is a measure of the steepness of the dose–response curve. A convenient way to measure n is to determine the IC_{50} and IC_{90} values from the dose–response curve and use the relationship:

$$n = \log 9/(\log[IC_{90}/IC_{50}]) \tag{6.2}$$

The relationship of Equation 6.1 assumes that the drug exposure lasts for some specified period of time. If the treatment duration changes, the relationship of Equation 5.32 can be used to correct the estimates of IC_{50} and IC_{90}.

Antibacterial therapy is probably the most straightforward of all therapeutic areas in which to do pharmacodynamic modeling. Dose–response relationships established *in vitro* can generally be applied directly to the *in vivo* situation, perhaps with appropriate correction for drug binding to plasma protein, as discussed in Chapter 3. If we know that a particular drug concentration, maintained for 24 h, gives 99.9% kill of a particular bacterial strain *in vitro*, there is a reasonable expectation that the same concentration will give a three log kill *in vivo*.

II. ANTIVIRAL MODELING

Antiviral chemotherapy is a level more difficult than antibacterial chemotherapy, for several reasons. Viruses are intracellular parasites that utilize many of the host cell reactions for their own purposes, so there are fewer pathogen-specific enzymes to exploit as selective targets. Since most of the existing antiviral target enzymes are part of the RNA or DNA replication machinery, many of the current antiviral drugs are nucleosides that require intracellular activation to the active nucleotide forms. Some DNA viruses and retroviruses are incorporated into the host DNA and may remain there for long periods as latent proviral nucleic acid that is essentially indistinguishable from the host cell's own genetic material. Models of the viral infection cycle and of antiviral drug action are thus, of necessity, more complex than the simple model of exponential bacterial growth described above. Circulating free virus must infect a susceptible host cell; this involves binding of the virus to an appropriate extracellular receptor of the host. Which receptor is used depends upon the virus; e.g., for the rhinoviruses that cause the common cold, the port of entry is the intercellular adhesion molecule (ICAM) receptor. Virus uptake is appropriately modeled as a saturable process. In practice, not all circulating virus particles (virions) are infectious, and not all virions have the opportunity to infect a host cell. Thus, a dimensionless parameter that is less than one, the **infectivity**, relates the number of circulating virions to the number of host cells infected in a particular time period. Once inside the host cell, the virion may be replicated manyfold (typically, perhaps 50-fold in 48 h), and the daughter virions are, in turn, released into the circulation. The virus infection and replication process will usually result in death of the infected host cell. In addition, infected cells may often display (in conjunction with MHC class 1 receptors) viral antigens, and this makes the infected cells targets for destruction by cytotoxic lymphocytes (killer T-cells). Latent infected cells do not normally release daughter virions, do not suffer virus-induced death, and probably are not targets for cytotoxic lymphocytes. Cytokinetic models of virus infections must describe these characteristic features of the process. However, many of the questions that arise in modeling antiviral therapy are similar to questions that we considered in connection with antibacterial therapy: what is the optimal schedule of administration? How much drug can we administer before host toxicity becomes a limiting factor? How do we minimize the development of drug resistance? (Discussion of this question is deferred to Chapter 7.) Does the host's immune system contribute to elimination of the infection? How can drugs be used most effectively in combination?

In general, basic information of antiviral activity and its spectrum, drug potency, and dose and time dependence can be obtained from *in vitro* measurements. Depending on the virus, a number of *in vitro* tests are used. A common test is the **plaque reduction assay**. This is used for those viruses that cause lysis of their host cells. A monolayer of the host cell line is grown in a Petri dish and infected by application to the plate of a very dilute suspension of the virus. Each virus particle will infect a cell, replicate, lyse its host cell, and in turn infect the surrounding cells. Thus, a zone of lysis, or plaque, will be seen corresponding to each virus in the original suspension. After a few days of culture, these plaques are typically about 2 mm in diameter. Now, if an antiviral drug is applied to the plate before the plaques develop, or if the virus suspension is treated with a drug before plating, the amount of virus killed will be reflected in a decrease in the number of plaques eventually seen on the plate. An IC_{50} concentration of inhibitor is then defined as that concentration that reduces the number of plaques by 50%, and an IC_{90} is the dose that lowers the number of plaques tenfold.

Another commonly used assay is the cytopathic effect (CPE) assay. For this assay a low-density cell culture (which may be a monolayer or suspension culture) is infected with a virus. After a few days, the number of host cells in the infected cultures, and in uninfected control cultures, are counted. Typically, the cell count in the infected culture will only be a few percent of the uninfected control value, because the virus has killed most of the infected cells. Now, if the test is carried out in the presence of an antiviral drug, the infected cell count will be higher, and for sufficiently high concentrations of an active drug, the infected cell count may approach or equal the uninfected control count (Figure 6.2). In the example shown, high concentrations of inhibitor (1 μM or greater) cause the final cell count of both infected and uninfected cultures to decline. This is because the test compound, at these concentrations, is causing direct cytotoxicity of the host cells. The antiviral activity is usually expressed as the median effective concentration, or EC_{50}, which brings the cell count up to half of the control value, or the EC_{90} concentration, which brings the count up to nine tenths of control. In the same assay, the cytotoxicity can be quantitated as an IC_{50}, the dose that brings the uninfected cell count down to 50% of the untreated value. The ratio of IC_{50} to EC_{50} then gives an *in vitro* estimate of therapeutic index. For a compound to stand a chance of being safe *in vivo*, the therapeutic range needs to be 100 or higher.

The ANTIBACT program package, discussed above (and in the Appendix) includes a related program, ANTIVIRL, that models the treatment of viral infections. Like ANTIBACT, ANTIVIRL makes it possible to explore the consequences of changes in doubling time, immunogenicity, and the effect of drug resistance on the efficacy of antiviral chemotherapy. It also factors in the effect of changes in virulence; it is sometimes observed that viruses acquire resistance to drugs at the expense of a decline in virulence. This parameter can be measured *in vitro*, and a cytokinetic model provides an approach to estimating how significant such changes are likely to be *in vivo*. Another difference between the ANTIVIRL model and models of bacterial proliferation is the added complication of latency, which can make viral infections impossible to cure, because latent provirus is out of the reach of drugs and of the immune system.

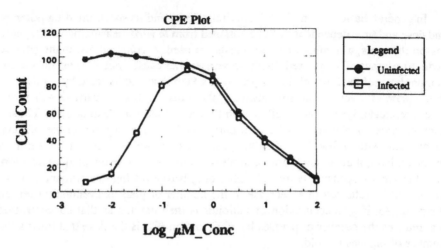

Figure 6.2   A cytopathic effect (CPE) assay for antiviral activity. The upper curve shows growth inhibition of uninfected cells at drug concentrations above 1 μ*M*. The lower curve shows an increase in the cell count at low inhibitor concentrations, due to antiviral activity, with toxicity to host cells again decreasing the cell count at higher concentrations.

III. MODELING AIDS INFECTIONS

Acquired immunodeficiency syndrome (AIDS) introduces yet another level of complication into chemotherapy and into our attempts to understand (and model) infections and their treatment. We have already considered the destruction of virally infected cells by the process of cellular immunity. The complication in AIDS is that its causative agent, the human immunodeficiency virus (HIV), uses cells of the immune system (a subset of T-cells, termed T-helper cells) as its host. Thus the virus destroys the immune system, while the immune system attempts to destroy the virus (or HIV-infected cells). This provides a closed loop that complicates the course of the disease and makes computer modeling essential to understanding the disease and its treatment. Another complicating factor in understanding the biology of HIV (or RNA viruses in general) is their very high mutation rate. The error rate for replication of RNA viruses is about 100-fold higher than for DNA replication, presumably because the cellular machinery for making RNA molecules lacks the sophisticated proofreading functions that keep DNA replications errors to a minimum. For HIV the error rate is about one per 10^4 bases. Since the HIV genome is about 10^4 bases, each daughter virion will differ from its parent on average at one locus. Thus a group of related virions do not form an identical clone, as is the case with mammalian cells, bacteria, or DNA viruses, but instead form a population of related but slightly different copies, referred to as a **quasispecies** (Eigen, 1993). This has two critical consequences for treatment of the disease. One is that drug resistance rates for HIV drugs are about 100-fold higher than for antibacterial drugs, a subject that is explored in Chapter 7. The other consequence is that HIV is unusually antigenically diverse. Nowack and May and their colleagues (Nowack

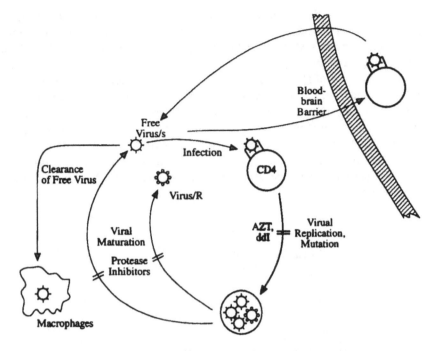

Figure 6.3 The HIV life cycle as modeled by the VIROCHEM program.

et al., 1991) suggest that over a period of years of continual HIV replication and turnover, less antigenic variants may arise (possibly because the capacity of the immune system to recognize new epitopes has been exhausted) that cannot be controlled by cytotoxic lymphocytes. When the course of an HIV infection passes this point, which Nowack and May refer to as the antigenic diversity threshold, the infection progresses to active AIDS.

A cytokinetic model, VIROCHEM, that describes these various aspects of HIV infections has been described (Jackson, 1996). Figure 6.3 summarizes those features of the HIV life cycle that are simulated by the model: the virus infects, and replicates within, CD4 cells. The production of DNA provirus strands may be inhibited by reverse transcriptase inhibitors, such as azidothymidine (AZT) and dideoxyinosine (ddI). Daughter copies of viral RNA are transcribed from the proviral DNA by RNA polymerase. Because reverse transcription and RNA polymerization are error-prone, mutant viruses are produced. Viral proteins are produced via a polyprotein that must be cleaved by HIV protease to make the mature proteins, and this process of viral maturation may be blocked by HIV protease inhibitors. Free viruses may be removed from the circulation by the macrophages of the reticuloendothelial system. HIV may cross the blood–brain barrier, replicate within the brain, and then cross back into the general circulation.

Probably part of the reason that AIDS has a long asymptomatic phase followed by sudden collapse of the immune system is that a reciprocal inhibition exists between the virus and the immune system (Figure 6.4). HIV infects and destroys CD4 cells, and because CD4 cells produce lymphokines required for replication of

CD8 cells (CTL), HIV also indirectly blocks CTL cell proliferation. However, CTL kills infected CD4 cells and thus limits HIV replication. This standoff between the virus and the immune system can last for many years. The reason it suddenly breaks down is a matter of debate, but, as noted above, the VIROCHEM model assumes that the explanation of Nowack and May is correct, and after the antigenic diversity threshold is reached, the infected cells become nonimmunogenic. The purpose of the model is to study questions concerning the relationship of pharmacokinetic properties to antiviral efficacy: what multiple of plasma concentration in relation to the *in vitro* antiviral level is it necessary to maintain? What kind of duration of response is it realistic to expect? How serious is missing an occasional dose likely to be? Does early treatment (before the antigenic diversity threshold is reached) give a major advantage?

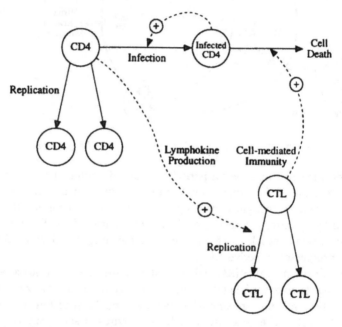

Figure 6.4 Interactions of HIV with the immune response.

The overall structure of the VIROCHEM model is shown in Figure 6.5. A pharmacokinetic module calculates plasma concentrations of drugs as a function of time. The drugs may be administered intravenously or orally, on any required schedule. For these drug levels, a pharmacodynamic module calculates the corresponding inhibition of virus replication, based upon experimentally determined dose–response curves. Next, a cytokinetics module that simulates the HIV life cycle and its interactions with the immune system predicts counts of CD4 and CD8 cells, and of wild type and mutant viruses in the various compartments. Table 6.1 shows the equations of this model; they include a PK expression, a set of mass action type equations describing the dose–response relationships of antiviral drug effects on sensitive and resistant viruses, and cytokinetic expressions for viral growth and CD4 and CD8 cell counts in the presence of inhibitors. Equation 6.5, which describes the inhibition of replication of drug-resistant viruses, was also used to describe the effect of reverse transcriptase (RT) inhibitors on bone marrow stem cells, with an appropriate value of the parameter R. Integration of this value then gives a measure of bone marrow toxicity. Finally, the incidence of drug-resistant mutants, at any particular total population size, is given by the Luria-Delbrück fluctuation equation. The parameter values used are given in the original publication (Jackson, 1996).

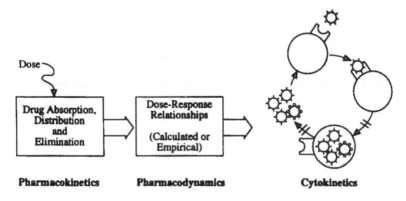

Figure 6.5   Block diagram of the VIROCHEM program.

One of the questions first studied with the VIROCHEM model was the issue of whether early treatment of HIV-positive subjects offered advantages over late treatment, or whether it was sometimes preferable to delay treatment. For a pathogen that is growing exponentially, if a given dose of drug provides a log kill of virus that is independent of viral load, then early or late treatment should give the same response duration (Figure 6.6). However, this simple analysis does not take into account the relative probabilities of drug-resistant viruses being present at low and high viral burdens. Moreover, because of the complex interactions between virus and immune system, the true growth curve is unlikely to be a simple exponential. Nowack et al. (1991) described a single simulation with their model comparing early and late treatment and predicted that early treatment should give a greater response duration. Their model described drug effects only in outline and did not consider the effects of toxicity of drug to the host. If the effects of host toxicity are factored in, the picture is somewhat more complex. Figure 6.7 summarizes simulations carried out using VIROCHEM. AZT causes cumulative toxicity to the host, which places a lifetime limit on how much can be administered. For late treatment (i.e., after the antigenic diversity threshold had been reached), the dose–response curve was biphasic, presumably because the drug effect on the infection is both dose and time dependent, but at higher doses the antiviral effect was nearing saturation, but the maximum cumulative dose was reached sooner, so that treatment had to be curtailed. With early stage treatment, AZT was assumed to be always given until the cumulative toxicity limit was reached. The response of early stage disease increased with dose up to a point, peaked at around an average IC_{50}, and then reached a plateau. The eventual virus breakthrough after early treatment with an RT inhibitor

Table 6.1 Equations of the VIROCHEM Model

1. Pharmacokinetics

$$C = F \cdot D/V_d \cdot [k_a/(k_a - k_e)] \cdot (e^{-k_e \cdot t} - e^{-k_a \cdot t}) \tag{6.3}$$

2. Dose–response relationships

(a)
$$a_s = 1/(1 + [C/IC_{50}]^{n_H}) \tag{6.4}$$

(b)
$$a_r = 1/(1 + [C/IC_{50}/R]^{n_H}) \tag{6.5}$$

(c)
$$n_H = \log 9/(\log[IC_{90}/IC_{50}]) \tag{6.6}$$

3. Cytokinetic expressions for viral growth in the presence of inhibitors

(a) Count of free virus/s

$$N_{s,t+1} = \underset{\text{maturation}}{N_{s,t} + I_s \cdot m_s \cdot a_{s2}} - \underset{\text{infection}}{N_{s,t} \cdot i \cdot H/(H+K_d)} - \underset{\text{turnover}}{N_{s,t} \cdot k_c} \tag{6.7}$$

Similar expressions for N_r and N_l (resistant and latent viral compartments)

(b) CD4 cell count

$$H = H_0 + \int_0^t \{\underset{\text{growth}}{(k_h H(A_h - \ln H))} - \underset{\text{infection}}{(N \cdot i \cdot a_{sl} \cdot H/(H+K_d))}\} dt \tag{6.8}$$

(c) Infected CD4 cell count

$$I_{s,t} = I_{s,0} + \int_0^t \{\underset{\text{infection}}{(N_{s,t} \cdot i \cdot a_{s1} \cdot H/(H+K_d))} - \underset{\text{lysis}}{I_{s,t} \cdot k_l} - \underset{\substack{\text{immune}\\\text{killing}}}{I_{s,t} \cdot T \cdot k_x}\} dt \tag{6.9}$$

Similar expression for I_r (CD4 cells infected with resistant virus)

(d) CD8 cell count

$$T = T_0 + \int_0^t \{(k_t T(A_t - \ln T)) \cdot V_{mh} \cdot H/(H+K_s)\} dt \tag{6.10}$$

4. Drug resistance

$$\mu = \alpha N (1 - N^{-\alpha-\beta})/(\alpha + \beta) \tag{6.11}$$

Table 6.1 (continued)   Equations of the VIROCHEM Model

Note:

α = mutation frequency

β = reversion frequency

μ = number of mutants

A_h = Gomperz asymptote parameter for CD4 cells

a_r = fractional activity of virus/R reverse transcription or polyprotein processing in the presence of inhibitor

a_s = fractional activity of virus/S reverse transcription or polyprotein processing in the presence of inhibitor.

   a_{s1}, a_{r1}: fractional activities resulting from effect of drug 1 (RT inhibitor) on sensitive and resistant virus, respectively

   a_{s2}, a_{r2}: fractional activities resulting from effect of drug 2 (protease inhibitor) on sensitive and resistant virus, respectively

A_t = Gomperz asymptote parameter for CD8 cells

C = concentration of inhibitor (nM)

D = drug dose (nmol)

F = fractional oral bioavailability

H = count of helper T-cells (CD4 cells); H_0 = zero-time count

i = infectivity of virus/s

IC_{50} = 50% inhibitory concentration of inhibitor for growth of virus/s

IC_{90} = 90% inhibitory concentration of inhibitor for growth of virus/s

I_r = number of CD4 cells infected with virus/R

I_s = number of CD4 cells infected with virus/S

$I_{s,t}$ = number of virus/s-infected CD4 cells at time t

K_a = affinity constant for lymphokine stimulation of CD8 cell replication

k_a = first order rate constant for oral drug absorption

k_c = first order rate constant for clearance of virus by macrophages

K_d = binding constant for attachment of virus to CD4 cells

k_e = first order rate constant for drug elimination = $\ln 2/t_{1/2}$

k_h = growth rate constant for CD4 cells = ln 2/doubling time

k_l = first order rate constant for virus-induced lysis of infected CD4 cells

k_t = growth rate constant for CD8 cells = ln 2/doubling time

k_x = second-order rate constant for cell-mediated cytotoxicity of infected CD4 cells

m_s = number of virus/s copies released per infected cell

N = total number of virus particles (virus/s + virus/R1 + virus/R2)

n_H = Hill coefficient for inhibition of maturation

N_r = number of free virus/R particles

N_s = number of free virus/S particles

$N_{s,t}$ = number of virus/S particles at time t

R = resistance factor = IC_{50}(virus/S)/IC_{50}(virus/R)

T = count of killer T-cells (CD8 cells); T_0 = zero-time count.

V_d = volume of distribution (l)

V_{mh} = constant of proportionality for lymphokine stimulation of CD8 cell replication

was always a drug sensitive virus, indicating that were it not for cumulative toxicity, more drug would have given a greater response. Early treatment could be either more or less effective than late treatment. At low doses, early treatment was more effective; at intermediate doses, late treatment was more effective; but at very high doses late treatment became less effective, probably because the treatment duration had to be decreased.

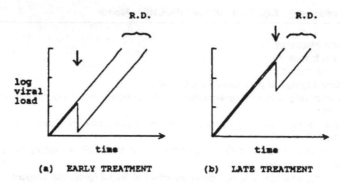

Figure 6.6 Growth curves for a virus that maintains exponential growth. Treatment with a drug dose that kills one log of virus (a) early and (b) late in the growth cycle. R.D., response duration.

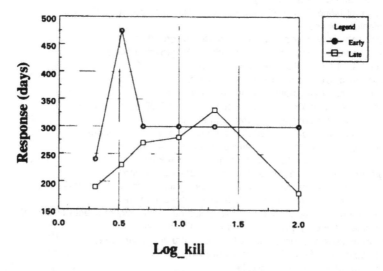

Figure 6.7 Predicted effects of early and late HIV infections with a hypothetical reverse transcriptase inhibitor, as calculated by the VIROCHEM program.

While it is still too early to know whether the class of HIV protease inhibitors will give cumulative toxicity after long-term treatment, the experience to date suggests that they do not, at any rate, cause the myelosuppression and peripheral neuropathy caused by the nucleoside RT inhibitors. For modeling purposes it was assumed that HIV protease inhibitors did not give long-term toxicity. This makes it possible to use high doses for (theoretically) unlimited periods of time, which means that the dose–response curve is monotonic. Initial simulations (Jackson, 1996) focused on the experimental protease inhibitor [3S-(3R±,4aR±,8aR±,2′S±,3′S±)] -2-[2′-hydroxy-3′-phenylthiomethyl-4′-aza-5′-oxo-5′-(2″-methyl-3″-hydroxyphe-nyl)pentyl]-decahydroisoquinoline-3-N-t-butyl carboxamide methanesulfonate (AG1343). Based upon allometric projections from preclinical data in four species (rat, dog, marmoset, and cynomolgus monkey) a dose of 400 mg, twice daily, of

AG1343 was predicted to maintain a sufficient plasma level to give a time-averaged 91.3% inhibition of HIV maturation. The results of simulations of AG1343 effects are summarized in Table 6.2. For a drug that has no cumulative toxicity, early treatment makes it possible to give a longer course of treatment, so when treatment is every day until treatment failure, it is perhaps not surprising that early treatment is superior to late treatment. However, even when the treatment of early disease is for the same length of time as the treatment of late disease, the model still predicts that early treatment should give better results.

Table 6.2 Predicted Effects of Early and Late Treatment of HIV Infections with the HIV Protease Inhibitor AG1343

Mean plasma level	IC_{50}	IC_{50}	IC_{80}	IC_{80}
Treatment duration (d)	620	3700	840	4150
Early treatment	420	950	680	1400
	(S)	(S)	(S)	(R2)
Late treatment	230	—	450	—
	(R2)		(R2)	

Note: Values shown are predicted response durations in days. (S) and (R2) indicate the predominant virus strain at the time of treatment failure. Calculations assume that virus/R2 is 120-fold resistant.

Another example of use of the VIROCHEM model to make development decisions about a development-stage anti-HIV drug (AG1343 again) concerned attempts to predict the likely decrease in circulating viral load. When AIDS patients are treated with an effective antiretroviral agent, the amount of HIV in their plasma, as measured by reverse PCR or by the branched-chain DNA technique, decreases. The circulating virus is known as a surrogate marker for the disease. Since HIV is believed to be the causative agent of AIDS, it seems reasonable that a decrease in the amount of virus in the blood should predict a positive therapeutic response. However, it has not yet been proven to the satisfaction of the FDA that this is the case. Because of the rapid turnover of HIV and of infected CD4 cells, the viral load reflects a balance between viral replication and turnover. The model predicted a direct relationship between mean drug levels and decrease in viral load: an average IC_{90} drug concentration gave a one-log decrease in viral load, an average IC_{99} drug concentration gave a two-log decrease in viral load, etc.

Figure 6.8 illustrates the predicted effect of AG1343 (300 mg, three times daily) on viral load. In this simulation of a late-stage HIV infection, at the time of treatment the circulating viral load was increasing rapidly and the CD4 count was falling. Following the initiation of treatment at day 374, viral load fell rapidly by over two logs. The CD4 count responded more slowly, but within 2 months of the initiation of treatment was projected to be approaching a normal count.

The above case studies provide examples of the use of cytokinetic modeling, linked to PK modeling, as a tool in drug development decision making. In the case of AG1343, the model made it possible to ask whether the PK and PD data obtained

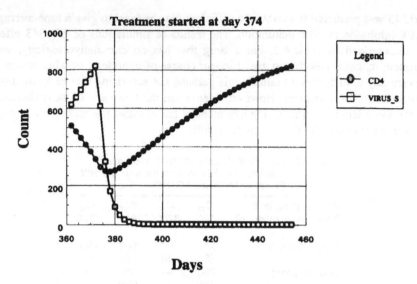

Figure 6.8 Calculated effects of an experimental HIV protease inhibitor on circulating viral load and on CD4 lymphocyte count.

from the preclinical studies were sufficiently encouraging to justify beginning expensive clinical trials. As the actual clinical data become available, they will provide a means of validating the predictions of the computer model.

IV. CYTOKINETICS OF MAMMALIAN CELLS

As with modeling of the cell kinetics of microorganisms, there are many questions concerning mammalian cell growth and its inhibition by drugs that may be profitably modeled at the whole-cell level. Some such models have described cell division and differentiation without detailed modeling of the cell cycle. Swan (1987) reviews the use of exponential, logistic, Gomperzian, and other growth expressions in describing tumor growth. Examples of the use of such models in describing drug effects are those of Jusko (1971) and of Skipper and Lloyd (Lloyd, 1977; Skipper, 1986). Other simple cytokinetic models have been used to simulate the process of differentiation, and these are discussed below. However, since most anticancer drugs cause perturbations of cell cycle progression, cytokinetic models of drug action usually attempt a description of the cell cycle. Jusko (1973) distinguished between S-phase and non-S-phase cells, and assumed that the former were sensitive to S-phase-specific drugs, and the latter were not. Ozawa et al. (1989) used this model to show that a representative non-phase-specific agent, cisplatin, gave cytotoxicity proportional to $C \times T$, while a cycle-specific agent, cytarabine, gave killing that was time dependent.

A large number of computer models have been written that describe drug effects on the cell cycle in some detail, though relatively few of them are available to the general user. An important example (because it is well documented and widely available) is CELLSIM (Donaghey, 1986), which has been implemented on the IBM-PC and other systems. CELLSIM can model the movement of cells between G_1, S,

G_2, M, and G_0 phases, and the time spent in these states may be set to any required value, which may be a constant or a probability distribution. The model can simulate results of cell cycle distribution studies measured by flow cytometry, so that predicted results can be compared with experimental data.

V. CELLULAR AUTOMATA AS CYTOKINETIC MODELS

The word *cellular*, used in this context, does not refer to cells in the biological sense. A cellular automaton refers to a pattern of repeating units, referred to as cells, that undergoes a series of transformations according to arbitrary rules. For example, the game of LIFE (Sigmund, 1993) takes place on a two-dimensional array of squares, each of which exists in one of two possible states (alive or dead, on or off, or whatever we choose to call them). At each successive move, each cell takes on a state that depends first, upon its current state and second, upon the state of its immediate neighbors. LIFE generates a series of patterns of "cells", some of which are stable (i.e., remain unchanged from one move to the next), others of which move in space, and yet others of which have the interesting ability to reproduce. As the name *LIFE* suggests, some of the properties of cellular automata are reminiscent of living systems. This, in turn, has led some investigators to attempt to explain living systems in terms of cellular automata. It is indeed possible to consider living cells as "cells" of cellular automata and their movement to different phases of the cell cycle or to quiescence or differentiation as "moves" that follow a system of rules (life as a game of LIFE). At present, this approach does not appear to be the most practical method of predicting drug effects, but the cellular automaton approach has some appealing features. Readers interested in reading about a cellular automaton model of tumor growth are referred to an article by Qu et al. (1993).

VI. PREDICTION OF CYTOSTASIS AND CELL DEATH: THE EXCHEMO MODEL

From the pragmatic perspective of a drug developer, the cytokinetic questions that are most likely to occur are along the following lines:

1. If compound A is known to be cell cycle specific, and we have measured its IC_{50} and IC_{90} against human colon carcinoma cells *in vitro*, and we know the achievable blood level and AUC in a mouse (and presumably have demonstrated *in vivo* activity against a human colon carcinoma line in a xenograft system), and if we predict the probable human pharmacokinetic properties (by allometry or from PB-PK modeling), how much tumor shrinkage are we likely to see from a maximum tolerated clinical dose (or course of treatment)?
2. If we now assume that (unlike the *in vitro* situation) 20% of the tumor cells are likely to be in G_0 phase at any time, but can be recruited out of it, how severely will this compromise the antitumor activity of compound A?
3. If the G_0 fraction is zero, but the tumor has a 30% loss factor, how will this affect sensitivity to compound A?

4. If the tumor doubling time *in vitro* was 24 h, but *in vivo* it slows down to 36 h, how is this likely to affect drug sensitivity?

5. If compound A causes cytotoxicity in S phase, what is the optimal way to combine it with a vinca alkaloid, which causes cell arrest and death in M phase?

The EXCHEMO program (Jackson, 1992; see Appendix) was written to study questions like this. EXCHEMO's view of the cell cycle is shown in Figure 6.9. Cells progress through the various phases of the cycle, or may spend time out of cycle in G_0. G_0 cells may be recruited back into cycle. Postmitotic cells may also leave the cycle irreversibly, to terminal differentiation, or to die (cell loss). Drugs may cause cell stasis or cell death at the indicated sites. The length of S, G_2, and M phases are treated as fixed, but the duration of G_1 may vary. The model simulates up to eight tumor cell populations: wild type (drug sensitive) cells, R1 cells (resistant to drug 1), R2 cells (resistant to drug 2), and R12 cells (resistant to both drugs). Each of these four subpopulations may be primary or metastatic. The mutation frequencies and degrees of drug resistance may be set to any required value. Resistant cells may back-mutate to drug sensitivity, and primary tumor cells may metastasize, but metastatic cells cannot revert to being primary tumor cells. In addition to the tumor cells, three normal cell populations are modeled: bone marrow, intestinal crypt cells, and T-lymphocytes. Drugs may be toxic to these normal populations, as well as to tumor cells. The T-cells may destroy tumor cells through cell-mediated immunity, and the degree of immunogenicity of tumor cells may vary from zero to highly immunogenic. Cell loss factor may be constant or it may be a function of tumor size (increasing in larger tumors). In the latter case, tumor growth approximates a Gomperzian growth curve. The fraction of tumor cells leaving the cell cycle to become quiescent (G_0) cells may be set to a constant or it may be treated as a function of tumor oxygen supply, as discussed below. In addition to modeling cytostatic and cytotoxic drugs, EXCHEMO can simulate effects of immunomodulators, differentiating agents, antimetastatics, and antiangiogenic drugs in any required combination. Drug concentrations may be constant (approximating the situation in tissue culture) or variable with time. The variable drug concentrations may be computed by a one-, two-, or three-compartment classical PK model or by a PB-PK model. In the latter case, the drug concentration–time profile experienced by primary and secondary tumors, bone marrow, and gut may be different, being calculated for the appropriate compartment. Thus, metastatic cells may be assumed to be in liver, brain, or any other site for which drug levels may be calculated.

Dose–response curves for EXCHEMO may be empirical or themselves the product of modeling calculations. In the former case they are obtained from *in vitro* measurements; in the latter case they may be calculated by any appropriate pharmacodynamic equation, including complex biochemical models. Thus, the EXCHEMO model combines PK modeling (classical or PB-PK) with pharmacodynamic models (empirical or biochemical) with cytokinetics. The resulting system makes it possible to simulate chemotherapy at three levels of complexity: the biochemical pathway, the cell, and the animal or patient. The use of this system is illustrated below by a number of examples (taken from Jackson, 1992).

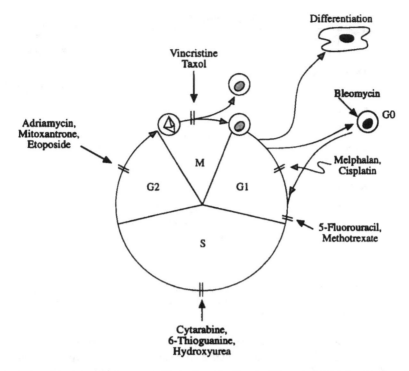

Figure 6.9   The EXCHEMO model of the cell cycle, showing sites of action of anticancer drugs.

A. EXCHEMO Case Study 1: The Importance of PK Parameters

Assume that we have two related cell cycle-specific anticancer drug analogues, one of which has a plasma elimination-phase half-life of 2 h, and the other, 15 h. They are otherwise identical. Which one should we develop? The simulations directed at this problem modeled murine tumors, but the principles would be the same for human tumors. Figure 6.10 shows the effect of a maximum tolerated dose of the drug with the 2-h half-life. The dose-limiting toxicity is to the gut, which loses two logs of crypt cells. The bone marrow stem cell population is also decreased by between one and two logs, reaches a nadir at about 30 h, then slowly starts to recover. A tumor with a 12-h doubling time (such as the L1210 leukemia) and an inoculum of 10^5 cells is totally eliminated in under 24 h under these conditions. By contrast, a tumor with a doubling time of 36 h (e.g., the Lewis lung carcinoma) reaches a nadir of about 25% of the original cell count at 32 h, then starts to recover (note that the recovering tumor cell population is partially synchronized). By 72 h, the tumor cell population is close to 70% of its pretreatment value. The results of this simulation were in agreement with the well-known experimental observation that cell cycle phase-specific drugs are much more active against rapidly dividing tumors than against slowly growing tumors.

Figure 6.11 shows the outcome of similar calculations carried out for a similar drug with a 15-h half-life. The dose-limiting toxicity (two log kill) is again to the

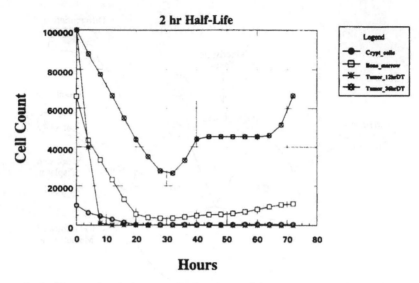

Figure 6.10   The predicted effects of a cell cycle-specific agent with a 2-h plasma half-life on
stem cells of bone marrow and intestinal crypts and on tumor cells with doubling
times of 12 h and 36 h.

gut. The bone marrow again suffers a loss of between one and two logs, reaches a
nadir (this time at 48 h) and slowly starts to recover. A tumor with 12-h doubling
time and inoculum of 10^5 cells is again eliminated within 24 h; the slowly cleared,
phase-specific drug, like its rapidly cleared analogue, is more active against a rapidly
dividing tumor than against a slowly dividing tumor. The interesting difference is
in the activity of the slowly cleared drug against the slowly dividing tumor: at 72 h
after treatment this tumor is still down at 13% of the pretreatment cell count. The
answer to the original question, is rapid or slow plasma clearance preferable? is that
it depends upon what the drug is for. To treat a tumor whose half-life is very short
(such as an acute leukemia), a drug with a long doubling time has no advantage and
is more toxic to normal cell compartments. To treat a tumor whose doubling time
is longer than that of bone marrow or intestinal crypt stem cells, a slowly cleared
drug, despite being more toxic, will have a therapeutic advantage. Of course, if the
drug is administered by continuous infusion or by frequent oral administration, the
PK differences become less marked. The likely effects of different administration
schedules can also be modeled, of course.

B. EXCHEMO Case Study 2: Optimal Timing of a Drug Combination

Imagine that we wish to treat the Lewis lung carcinoma of mice with a combination of vincristine and cytarabine. The Lewis lung carcinoma has a relatively long
doubling time for a transplanted murine tumor (36 h), and vincristine and cytarabine
are both cell cycle-specific agents, so we do not expect a strong therapeutic effect
in this system with these drugs. Cytarabine blocks and kills cells in S phase, and
vincristine causes M-phase arrest. At low, nonlethal doses, vincristine tends to
synchronize cells, so the question arises, can we time the two drugs so that the

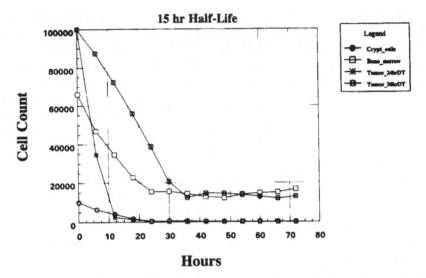

Figure 6.11 The predicted effects of a cell cycle-specific agent with a 15-h plasma half-life on stem cells of bone marrow and intestinal crypts and on tumor cells with doubling times of 12 h and 36 h.

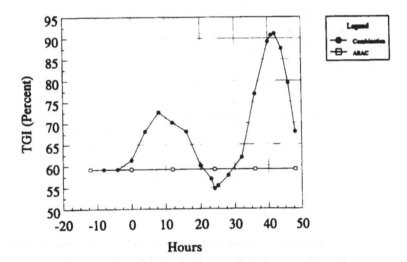

Figure 6.12 Use of the EXCHEMO program to predict the antileukemic effect of combinations of vincristine and cytarabine, showing the importance of optimal scheduling.

blocked cells move synchronously into S phase, and the cytarabine is administered when the number of cells in S phase is at its peak? This kind of trick works very well in tissue culture, where drugs can be removed from the medium essentially instantaneously; it is much more difficult to synchronize tumor cells *in vivo*, where the vincristine stays around inside the cells for several hours. Figure 6.12 shows the results of a series of simulations carried out with EXCHEMO. We assume that the dose of vincristine is low (1 mg/kg, using a three-compartment classical PK model) since we want it to synchronize cells, not to kill them. A therapeutic, but nontoxic (to the host), dose of cytarabine (10 mg/kg, using a one-compartment PK model) is administered at times varying from 12 h before the vincristine to 48 h after the vincristine. Tumor growth inhibition, or TGI, is assessed 72 h after the start of treatment. TGI is the treated tumor size as percent of an untreated control substated from 100. Vincristine alone, at this low dose, gave TGI of 4.4%, and cytarabine alone gave a modest TGI of 59.2%. Depending upon the timing, the combination was worse than, equivalent to, or better than cytarabine alone. When cytarabine was administered 12, 8, or 4 h before vincristine, the combined effect was identical to that of cytarabine alone. Simultaneous administration was slightly better than cytarabine alone, and delaying the cytarabine administration further gave an even better result. The combined effect peaked when the delay between vincristine and cytarabine was 8 h. Between 22 and 28 h delay, the combined effect was worse than cytarabine alone, because the synchronized cohort of cells was moving through G_2, M, and G_1 phases, where cytarabine is almost inactive. With a delay between vincristine and cytarabine of 32 h or more, the combined effect became much greater, and it peaked at 42 h, with a TGI of greater than 90% (still modest, but satisfactory for two antileukemia drugs acting on a relatively slowly growing solid tumor). Note that the two peaks of activity were separated by 36 h, the cell cycle time. Of course, the optimal timing would be different for a tumor with a different cell cycle time. There have been numerous experimental studies showing that partial *in vivo* synchronization can sensitize tumors to cell cycle phase-specific cytotoxic agents. There is more controversy about whether this is a useful strategy in humans, mainly because human tumors are harder to synchronize than transplanted murine tumors, which have more heterogeneity of cytokinetic parameters, including cell cycle time.

Clearly, to simulate this kind of therapeutic approach requires a model that describes the cell cycle, and the interactions of drugs with it, in some detail. A program like SKIPPER (which does not model the cell cycle) incorrectly predicts a similar outcome for the various sequences of administration of the two drugs. In contrast, EXCHEMO, which models PK and cytokinetic effects, is able to predict the outcome of such experiments with considerable accuracy.

C. EXCHEMO Case Study 3: Optimal Use of Antimetastatic Agents

There is much literature on the various stages of metastasis — migration of cells from the primary tumor, intravasation into a capillary, dissemination through the vascular system, fibrin clot formation and attachment of the clot to a capillary wall, extravasation of metastatic cells from the capillary, and, finally, replication of the matastatic cell to form a secondary tumor. Compounds are known that inhibit each of these steps, but few, if any, antimetastatic agents that have been taken to clinical trial have shown convincing activity. The usual argument is that if a tumor has not metastasized at the time of diagnosis, early surgery is likely to be curative, but that if the tumor has already metastasized at diagnosis, it is too late for antimetastatic agents to influence the course of the disease. It is unlikely that drug developers will spend large sums of money on clinical trials of new antimetastatic drugs unless this kind of objection can be convincingly countered. If an antimetastatic agent has sufficiently low toxicity, it could be tested as a preventive agent in certain groups known to be at very high risk for cancer. With established and inoperable tumors, an antimetastatic drug might at least delay and decrease the establishment of additional metastases. By the time treatable tumors are diagnosed, they are probably shedding metastatic cells faster than at any time in their previously occult life history; the immediate institution of antimetastatic treatment, before surgery or radiation could be used, might greatly limit the total number of metastases. Finally, it has been suggested that some surgical techniques may themselves encourage dissemination of metastatic cells; if so, the use of an antimetastatic drug (preferably one that does not inhibit blood clotting) before surgery might prevent this. These are the kinds of situations where simulation can explore the various treatment and prevention strategies and suggest approaches where an antimetastatic may make a difference and where it is likely to be a waste of time.

Table 6.3 summarizes some simulations run with EXCHEMO. Again, a mouse tumor was modeled, with a doubling time of 24 h assumed for both primary and metastatic cells. The tumor is highly metastatic, and after 48 h of uninhibited growth, over 11,000 metastatic cells are present (Table 6.3, top). If the primary tumor is removed at 24 h (Table 6.3, bottom) the final number of metastatic cells is reduced by 50%; this number nevertheless doubles between 24 and 48 h. The number of secondary cancer cells in a tumor-bearing animal increases for two reasons: replication of cells already present in the metastatic foci and migration of metastatic cells out from the primary lesion. Surgery, of course, will stop the second source, but not the first.

Table 6.4 (bottom) shows the effect of the cyclooxygenase inhibitor, indomethacin, on accumulation of metastatic cells in this system. Indomethacin has no effect on the growth of the primary tumor, but the number of metastatic cells at both 24 h and 48 h is inhibited by 82% relative to the untreated control. If the indomethacin is given at 24 h only, the effect on the secondary tumor cell population resembles that of surgery at 24 h. Table 6.5 shows the effect of combined surgery and antimetastatic therapy, surgical removal of primary tumor again being performed at 24 h. The top panel shows the result of starting indomethacin at 24 h; the result is identical to that of surgery alone. Because the primary tumor has been excised, no new metastatic cells are being shed, and because antimetastatic agents do not affect the growth of existing secondary tumors, there is no advantage in using an antimetastatic drug in this situation. Table 6.5 (bottom) shows the effect of starting indomethacin at zero time, in combination with surgery at 24 h. This combined modality treatment resulted in over 90% suppression of metastatic cells at 24 h.

D. EXCHEMO Case Study 4: Immunogenicity and Drug Response

The existence of an immune response directed against tumor cells has been clearly demonstrated in rodents, and there is reason to believe that at least some human tumors evoke a cell-mediated immune response. Cytotoxic T-lymphocytes,

Table 6.3   Cytokinetic Simulation of a Metastatic Tumor

EXCHEMO: Control Growth of a Metastatic Tumor
Tumor Doubling Time = 24 h   MC = 422

Time	Tumor cells	Met. cells	Crypt cell s	Bone marrow
0	1,000,000	0	10,008	66,000
4	1,122,248	242	10,008	66,000
8	1,259,500	510	10,008	66,000
12	1,413,945	870	10,010	66,001
16	1,586,811	1,454	10,010	66,003
20	1,780,967	2,114	10,010	66,003
24	1,998,846	2,860	10,010	66,003
28	2,243,245	3,697	10,010	66,003
32	2,517,616	4,635	10,010	66,003
36	2,826,268	5,816	10,010	66,003
40	3,171,814	7,473	10,010	66,003
44	3,559,888	9,360	10,010	66,003
48	3,995,379	11,485	10,010	66,003

Tumor growth inhibition = –0.2%
Gross log tumor cell kill = 0
Net log tumor cell kill = –0.6

EXCHEMO: Growth of a Metastatic Tumor with Surgical Excision of Primary
Tumor Doubling Time = 24 h   MC = 422

Time	Tumor cells	Met. cells	Crypt cell s	Bone marrow
0	1,000,000	0	10,008	66,000
4	1,122,248	242	10,008	66,000
8	1,259,500	510	10,008	66,000
12	1,413,945	870	10,010	66,001
16	1,586,811	1,454	10,010	66,003
20	1,780,967	2,114	10,010	66,003
24	1,998,846	2,860	10,010	66,003

Surgical removal of primary tumor at 24 h

Time	Tumor cells	Met. cells	Crypt cell s	Bone marrow
28	0	3,211	10,010	66,003
32	0	3,605	10,010	66,003
36	0	4,060	10,010	66,003
40	0	4,539	10,010	66,003
44	0	5,092	10,010	66,003
48	0	5,718	10,010	66,003

Tumor growth inhibition = 99.9%
Gross log tumor cell kill = 2.84
Net log tumor cell kill = 2.24

natural killer cells, and activated macrophages all contribute to varying degrees to antitumor immunity. With rare exceptions, the immune response to human tumors does not result in spontaneous remission of disease, but it is a mistake to conclude that

Table 6.4 Simulation of the Effect of an Antimetastatic Drug

EXCHEMO: Indomethacin (8 mg/kg) at 24 h
Tumor Doubling Time = 24 h   **MC = 422**

Time	Tumor cells	Met. cells	Crypt cell s	Bone marrow
0	1,000,000	0	10,008	66,000
4	1,122,248	242	10,008	66,000
8	1,259,500	510	10,008	66,000
12	1,413,945	870	10,010	66,001
16	1,586,811	1,454	10,010	66,003
20	1,780,967	2,114	10,010	66,003
24	1,998,846	2,860	10,010	66,003
28	2,243,245	3,216	10,010	66,003
32	2,517,616	3,630	10,010	66,003
36	2,826,268	4,141	10,010	66,003
40	3,171,814	4,758	10,010	66,003
44	3,559,888	5,607	10,010	66,003
48	3,995,379	6,783	10,010	66,003

Tumor growth inhibition = −0.1%
Gross log tumor cell kill = 0
Net log tumor cell kill = −0.6

EXCHEMO: Indomethacil (8 mg/kg) at 0 and 24 hr
Tumor doubling time = 24 h   **MC = 422**

Time	Tumor cells	Met. cells	Crypt cell s	Bone marrow
0	1,000,000	0	10,008	66,000
4	1,122,248	2	10,008	66,000
8	1,259,500	10	10,008	66,000
12	1,413,945	36	10,010	66,001
16	1,586,811	100	10,010	66,003
20	1,780,967	244	10,010	66,003
24	1,998,846	514	10,010	66,003
28	2,243,245	596	10,010	66,003
32	2,517,616	770	10,010	66,003
36	2,826,268	1,015	10,010	66,003
40	3,171,814	1,159	10,010	66,003
44	3,559,888	1,479	10,010	66,003
48	3,995,379	2,092	10,010	66,003

Tumor growth inhibition = 0.1%
Gross log tumor cell kill = 0
Net log tumor cell kill = −0.6

it is unimportant. This point is illustrated by the following comparison of two murine tumors, identical except that one is nonimmunogenic and the other is moderately immunogenic. Table 6.6 (bottom) shows the effect of a maximum tolerated dose of cisplatin on a rapidly dividing murine tumor. This dose gives an almost two-log decrease in the intestinal crypt stem cell population, which reaches a nadir after

Table 6.5   Use of an Antimetastatic Drug Combined with Surgery

EXCHEMO: Surgery and Indomethacin at 24 h
Tumor Doubling Time = 24 h   **MC = 422**

Time	Tumor cells	Met. cells	Crypt cells	Bone marrow
0	1,000,000	0	10,008	66,000
4	1,122,248	242	10,008	66,000
8	1,259,500	510	10,008	66,000
12	1,413,945	870	10,010	66,001
16	1,586,811	1,454	10,010	66,003
20	1,780,967	2,114	10,010	66,003
24	1,998,846	2,860	10,010	66,003

Surgical removal of primary tumor at 24 h

28	0	3,211	10,010	66,003
32	0	3,605	10,010	66,003
36	0	4,060	10,010	66,003
40	0	4,539	10,010	66,003
44	0	5,092	10,010	66,003
48	0	5,718	10,010	66,003

Tumor growth inhibition = 99.9%
Gross log tumor cell kill = 2.84
Net log tumor cell kill = 2.24

EXCHEMO: Indomethacin at 0 h and Surgery at 24 h
Tumor Doubling Time = 24 h   **MC = 422**

Time	Tumor cells	Met. cells	Crypt cells	Bone marrow
0	1,000,000	0	10,008	66,000
4	1,122,248	2	10,008	66,000
8	1,259,500	10	10,008	66,000
12	1,413,945	36	10,010	66,001
16	1,586,811	100	10,010	66,003
20	1,780,967	244	10,010	66,003
24	1,998,846	514	10,010	66,003

Surgical removal of primary tumor at 24 h

28	0	591	10,010	66,003
32	0	745	10,010	66,003
36	0	935	10,010	66,003
40	0	942	10,010	66,003
44	0	963	10,010	66,003
48	0	1,028	10,010	66,003

Tumor growth inhibition = 100%
Gross log tumor cell kill = 3.59
Net log tumor cell kill = 2.99

16 h, and then starts to recover. The bone marrow stem cell population is decreased by about 70%. The tumor cell population is reduced by over 2.8 log by this very effective agent, but after 12 h the tumor begins to recover. Table 6.6 (top) shows the effect of the same treatment regimen on a moderately immunogenic tumor; the

Table 6.6   The Effect of Immunogenicity
                 on Drug Response

EXCHEMO: Moderately Immunogenic Tumor
Tumor Doubling Time = 12 h   **MC = 412**
Cisplatin = 15 mg/kg (0 h)

Time	Tumor cells	Crypt cells	Bone marrow
0	200,000	10,008	66,000
4	2,070	488	21,663
8	599	253	20,119
12	379	155	21,175
16	545	111	24,054
20	343	115	27,254
24	180	134	30,471
28	200	206	34,919
32	40	218	37,984
33	0	218	38,531

Tumor eradicated

Tumor growth inhibition = 100%
Gross log tumor cell kill = 6.13
Net log tumor cell kill = 5.3

EXCHEMO: Nonimmunogenic Tumor
Tumor Doubling Time = 12 h   **MC = 412**
Cisplatin = 15 mg/kg (0 h)

Time	Tumor cells	Crypt cells	Bone marrow
0	200,000	10,008	66,000
4	2,143	488	21,663
8	761	253	20,119
12	657	155	21,175
16	1,066	111	24,054
20	1,117	115	27,254
24	1,110	134	30,471
28	1,878	206	34,919
32	2,116	218	37,984
36	2,282	218	40,983
40	3,688	218	45,352
44	4,229	228	49,061
48	4,673	270	52,598

Tumor growth inhibition = 99.9%
Gross log tumor cell kill = 2.84
Net log tumor cell kill = 1.63

effects of the drug on the normal tissues are identical, but in this case the tumor is completely eliminated by 33 h. The growth rates of the immunogenic and nonimmunogenic tumors in the absence of drug are almost identical (not shown), but, as shown by these two simulations, immunogenicity can make a difference between a curative and subcurative response to drug, at least for a small tumor.

E. EXCHEMO Case Study 5: The Problem of Quiescent Tumor Cells

Cell cycle phase-specific anticancer drugs are well known to be most active against tumors with a high growth fraction. Indeed, these drugs derive part of their selectivity from the fact that the bone marrow stem cell population contains a proportion of cells (G_0 or Q cells) that are not in cycle at any particular time, but which retain the ability to be recruited back into the cycle. Clearly this advantage is lost if the tumor also contains a quiescent fraction that can be recruited back into cycle. Table 6.7 (top) shows the curative treatment of a rapidly growing murine tumor, whose growth fraction is 100%, by the antileukemia agent cytarabine. Table 6.7 (bottom) shows the effect of a similar treatment on an otherwise identical tumor that contains a quiescent fraction of 1%. In this case the tumor cell population reaches a nadir at 28 h, at which time almost two logs of tumor cells have been killed, and then starts to recover. Thus, a very small quiescent fraction has had a profound effect on the drug responsiveness of the tumor. The quiescent fraction could be destroyed by prolonging the cytarabine treatment, to pick off the Q-cells as they are recruited back into the cell cycle; the trouble with this approach is that it will also greatly increase toxicity to bone marrow. If the tumor Q fraction is smaller than that of the bone marrow, this approach may still help; if we know enough about the cytokinetics of the tumor, modeling approaches can help to optimize the killing of quiescent tumor cells while keeping toxicity to normal tissues at a tolerable level.

F. EXCHEMO Case Study 6: Modeling Tumor Angiogenesis

Solid tumors generally have a limited and poorly organized blood supply. It is believed that tumors cannot grow to more than about a millimeter in diameter without inducing the growth of new blood vessels into them, a process known as neoangiogenesis. Without this, the tumor becomes hypoxic, the quiescent fraction increases, and cells near the center of the growth become necrotic. The angiogenesis process involves production, by the tumor cell, of growth factors that stimulate the proliferation and motility of nearby capillary endothelial cells, which are thus induced to invade the growing tumor. A recent approach to inhibiting tumor growth is the development of antiangiogenic agents, which block the process of new blood vessel formation. There is some debate about how effective such drugs are likely to be as single agents, since presumably the outer layer of tumor cells should be adequately oxygenated, even in the absence of a blood supply, which means the tumor should be able to continue growth, if not exponentially, at least linearly. As with antimetastatic agents, before spending a lot on clinical development of an antiangiogenic, we may want to contemplate how such an agent could be used most effectively, and a cytokinetic model can help us to do this.

Figure 6.13 shows the result of an EXCHEMO simulation of growth of a tumor from an initial size of 10^7 cells. The tumor was assumed to be autochthonous, i.e., spontaneously arising, rather than transplanted; thus, at the start of the study the tumor already had an established blood supply. The control tumor grew logarithmically until a lethal tumor size was reached, indicating that its blood supply kept up with its growth. The "treated" line on Figure 6.13 shows the predicted growth of a

Table 6.7 The Effect of Quiescent Cells
on Drug Response

EXCHEMO: No Q Fraction
Tumor Doubling Time = 12 h   MC = 212
Cytarabine = 10 mg/kg (0 h)

Time	Tumor cells	Crypt cells	Bone marrow
0	1,000,000	10,008	66,000
2	495,320	7,105	48,541
4	401,479	6,281	43,529
6	154,373	5,467	38,535
8	13,551	4,658	33,545
10	218	3,866	28,605
12	7	3,102	23,756
14	0	2,378	19,062

Tumor eradicated

Tumor growth inhibition = 100%
Gross log tumor cell kill = 6.35
Net log tumor cell kill = 6

EXCHEMO: Q Fraction of 1%
Tumor Doubling Time = 12.1 h   MC = 212
Cytarabine = 10 mg/kg (0 h)

Time	Tumor cells	Crypt cells	Bone marrow
0	1,000,000	10,008	66,000
2	502,095	7,105	48,541
4	411,116	6,281	43,529
6	168,509	5,467	38,535
8	30,017	4,658	33,545
10	16,298	3,866	28,605
12	15,356	3,102	23,756
14	14,637	2,378	19,062
16	13,965	1,709	14,604
18	13,350	1,120	10,541
20	12,804	726	7,735
22	12,347	566	6,453
24	11,996	564	6,019
26	11,711	696	6,180
28	11,614	914	6,922
30	11,697	986	7,346
32	11,940	986	7,594
34	12,308	986	7,919
36	12,801	983	8,282
38	13,441	977	8,629
40	14,272	968	8,934
42	15,325	962	9,198
44	16,591	962	9,429
46	18,030	980	9,678
48	19,652	1,086	10,123

Tumor growth inhibition = 99.9%
Gross log tumor cell kill = 2.9
Net log tumor cell kill = 1.71

similar tumor in mice treated daily with a corticosteroid agent at a dose sufficient to block angiogenesis completely. While the tumor was small, the treated group grew at the same rate as the untreated controls, showing that tumor blood supply was not limiting. After the size reached about 7.6 logs of tumor cells, however, its growth rate slowed down, and the model predicted that the tumor now contained a high fraction of quiescent cells. Despite this, the treated tumor continued to grow and eventually reached a lethal size. By the end of the study, the growth fraction of the tumor was only about 0.06. If a mouse bearing a transplanted tumor is treated with antiangiogenic agents right from the time of implantation, the slowing of tumor growth is more dramatic; however, this is a less clinically relevant situation than the autochthonous tumor modeled in Figure 6.13. These predictions suggest that anti-angiogenic agents, used alone, may be insufficient to cause total arrest of the growth of an established tumor.

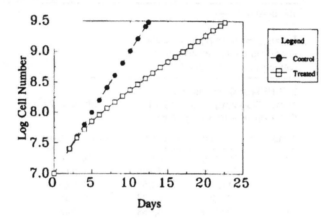

Figure 6.13 Use of the EXCHEMO program to model the effect of a corticosteroid antiangiogenic agent on growth of a murine solid tumor.

G. Normal and Tumor Cell Differentiation

Some types of tumors have a tendency to differentiate spontaneously into cells with limited (or no) capacity for further replication. For such tumors the probability of a cell undergoing differentiation is proportional to the amount of time it spends in G_1 phase; thus drugs that cause G_1 arrest tend to induce differentiation in these tumors. In addition to G_1-blocking agents, differentiation may be caused by various noncytotoxic compounds, such as nonpolar solvents (e.g., dimethylformamide). The EXCHEMO model can simulate both spontaneous and induced differentiation, and it may thus be used to study questions concerning the feasibility of using differentiating agents as a treatment approach for human cancer. However, the EXCHEMO program does not permit modeling the details of normal cell differentiation, and the author is not aware of any commercial modeling programs written specifically for this purpose. There are several articles in the literature describing modeling of the differentiation of hematopoietic precursor cells, and a reader experienced in dynamic modeling could no doubt reproduce these models with a little effort. Blumenson

(1975) described a detailed model of the development of granulocyte precursors in humans. The model describes stem cell kinetics and maturation through myeloblasts, promyelocytes, early and late myelocytes, metamyelocytes, and band and segmented neutrophils. The phases of the cell cycle are modeled, so that, for example, the effects of an S-phase-specific drug can be simulated. Feedback in the system allows the early precursors to respond to a loss of circulating neutrophils (as a result of blood loss, for example) by recruiting quiescent precursor cells into the cell cycle. The publication gives sufficient data, including normal parameter values for a 70-kg human, to allow readers to reproduce the model. Blumenson goes on to describe a detailed modeling study of the effects of 5-fluorouracil on circulating neutrophils and on every stage of precursor cells. The model's predictions agree well with clinical observation, showing a nadir at 10 d followed by an overshoot at 20 d. Because neutropenia is such a common side effect of cancer chemotherapy, this detailed model of granulopoiesis is a useful tool for understanding myelotoxicity in humans.

Düchting (1978) has provided a similarly detailed model of erythropoiesis. His model contains nine compartments, ranging from the pluripotent myeloid stem cell to the mature red blood cell. Düchting created his model using a general-purpose simulation package called ASIM, created by the German engineering firm AEG-Telefunken. The author has no information on the current availability of this simulation program; presumably, the results could be reproduced with a modern PC-based package, such as STELLA II (see Appendix). Düchting's model described the stimulating effect of erythropoietin and also responded to such external factors as blood loss (by bleeding or by drug effects), dietary iron deficiency, and tissue oxygen levels. The kinetics of a number of hematological diseases can be described with this model, and in some cases they are interpreted as caused by instability of feedback control loops.

H. Models of the Immune Response

The control of the immune response is highly complex, involving numerous cell populations and cytokines and multiple feedback loops, so that regulatory effects are essentially impossible to keep track of without the help of a computer model. Since an understanding of immune function is necessary to predict the activity of many classes of drugs, including anti-infective agents, anticancers, immunosuppressives, asthma drugs, and agents for autoimmune diseases (as well as vaccines), there is probably no area where the need for a good computer model is greater. To the author's knowledge, a definitive model that encapsulates present knowledge about the control of the immune system and the effects of known drugs on that system does not yet exist.

A number of authors have modeled the interactions between B-cells, cytotoxic T-cells, helper T-cells, suppressor T-cells, NK-cells, macrophages, and the various lymphokines and the effects of these components of the immune system upon tumor cells. A review by Dullens et al. (1986) described the models developed at that time that concentrated on describing tumor cell rejection, with particular emphasis on immune surveillance against neoplasia, factors that allow tumors to escape from immune attack, and interactions between macrophages and T-lymphocytes that generate a cellular antitumor immune response.

What drugs developers need is a model into which observed *in vitro* effects can be factored in an attempt to predict the possible *in vivo* consequences; e.g., if a new compound decreases IL-2 production by 50% in cultured T-cells, but doubles the level of TNFα production by macrophages, is it likely to increase or suppress an antitumor immune response? In this, as in many other areas of pharmacodynamics, the future will bring off-the-shelf software that will help the nonexpert not only to access the current knowledge, but to manipulate it, and predict the probable *in vivo* effects of unstudied compounds. In the meantime, the best interim strategy is for users to use general-purpose simulation packages (SIMUSOLV and STELLA II) to create small, special-purpose models focused on specific pharmacodynamic questions.

CHAPTER 7

Prediction of Drug Resistance

Consider the situation of a person recently infected with human immunodeficiency virus. On average, it is likely that 7 to 10 years will elapse before the infection progresses to AIDS. Thus, if we could treat infected individuals, as soon as they seroconvert, with a drug that gives 90% inhibition of viral replication, then the disease, while still incurable, should be staved off for at least 70 years. A number of drugs are available that reach and maintain an average IC_{90} concentration in the blood. Since, unfortunately, we cannot sustain 70-year responses with these drugs, we must ask where the fallacy is in our original argument. In fact, control of AIDS (and of many other hard-to-treat infections and tumors) is limited by two factors: cumulative drug toxicity and acquired drug resistance. The first of these factors means that there is a limit to how much drug we can administer, and for how long; the second factor means that the disease anyway ceases to respond after a while. Drug companies that are in the business of developing antibiotics are familiar with the resistance problem, and one of their rules of thumb (heuristics) says that you have, on average, 8 to 10 years to recoup development costs before drug resistance markedly limits the usefulness of your new antibiotic (at least against serious hospital-transmitted infections). When we now consider that the mutation rate of HIV is about 100-fold greater than for *Staphyllococcus aureus*, this suggests that an AIDS drug will have a useful life of about 5 weeks (assuming the relationship is a linear one). No company is likely to spend $100 million developing a drug with such a short product cycle, and some companies have already allowed this argument to keep them out of the AIDS drug development arena. The companies that have remained in the area presumably do not buy this argument, but worry about it nevertheless. This argument is revisited (with nonlinear mathematics) below.

Drug resistance is a particular problem of chemotherapy, both antimicrobial and anticancer chemotherapy. Hypertension patients do not become resistant to their beta-blockers, nor do diabetics become resistant to insulin (there is a condition called insulin-resistant diabetes, but that is a different disease from the insulin-responsive kind). In the early days of chemotherapy there was a lengthy debate about whether drug resistance was genetic or whether it was a kind of epigenetic adaptation to the continued presence of drug. It is now clear that drug resistance is a heritable genetic change, which may be caused by point mutations or by other chromosomal changes, such as deletions, translocations, or gene amplification. The mutations are stable and tend to persist in the absence of drug, though over a period of time in the absence of the selecting agent, reversion to the drug sensitive phenotype does tend to occur. Resistance to a particular drug may be caused by multiple mechanisms, and, conversely, a single mutation may confer resistance to multiple drugs.

A knowledge of likely mechanisms of drug resistance, of the probable degree and frequency of resistance, and of cross-resistance patterns with other drugs is an essential part of the preclinical workup for a new chemotherapeutic agent. What is not routinely done at present is to factor the information about resistance into clinical trial design, because this requires some fairly sophisticated modeling capability.

Modeling the development of drug resistance is a branch of cytokinetics, since it must be understood at the cellular level. The current discussion is thus a reprise of Chapter 6, but with the emphasis now on the single aspect of drug resistance. Before getting down to details, it is necessary to clarify two points of terminology. The first is the question of **intrinsic resistance** vs. **acquired resistance**. Intrinsic (or primary) resistance is where a particular infection (or tumor) fails from the outset to respond to a treatment protocol that is known to be effective against other infections (or tumors) of a similar kind. Acquired (or secondary) resistance is where an infection or tumor that initially responded to a particular treatment recurs in a form that no longer responds to that treatment. There has been much discussion of whether the same biochemical factors are involved in the two phenomena. In some cases, at least, the biochemical mechanisms of intrinsic and acquired resistance to a particular drug do appear to be similar. The difference lies in the dynamics of response of the disease. An intrinsically resistant disease simply fails to respond from the beginning of treatment; the cells that kill the animal or patient are the same kind of cells that we originally attempted to treat. With acquired resistance, the cells that kill the host animal are biochemically altered (mutant) cells that no longer respond to the drug that was effective against the original disease. Acquired resistance is a phenomenon of genetic diversity, of genetic mutation.

The other frequent question that crops up, which is partly one of nomenclature, is the issue of whether one should distinguish between resistant mutants that emerge during the course of treatment and those that were present, from the beginning, as a resistant subpopulation. Consider a population of bacteria that mutate to a penicillin-resistant state with a frequency of one per 10^6 organisms. If treatment is started when the total infection consists of 10^4 organisms, it is most unlikely that any resistant mutants are present. If we undertreat the infection, so that it manages to grow up to a population of 10^7 bacteria, the chances are that there will be resistant organisms present by that time. If there were 10^7 bacteria when treatment began, there would be resistant bacteria present from the beginning. Biochemically, the mutants would be the same in both cases. Both populations follow the same mathematical relationships, and both are examples of acquired resistance. In fact, the first population of 10^7 cells (the one that had started out at 10^4 cells) would contain more (probably nearly all) resistant cells, because it had been subjected to selective pressure (from the unsuccessful treatment) against the sensitive cells.

I. THE FLUCTUATION EQUATION

The fundamental equation of drug resistance was described in 1943 by Luria and Delbrück, and is known as the **fluctuation equation**:

$$\mu = \alpha N\,(1 - N^{-\alpha-\beta})/(\alpha + \beta) \tag{7.1}$$

where μ is the number of mutant cells in a total population of N cells, α is the mutation rate, and β is the frequency of back-mutation from resistance to sensitivity. μ is the median number of mutants that would be seen in a number of populations of N organisms. Because mutation appears to be a random event, the actual number of mutants will vary and will approximate a Poisson distribution:

$$P_\mu = [e^{-\alpha N}\,(\alpha N)^\mu]/\mu! \tag{7.2}$$

where P_μ is the probability of μ mutants being seen in the N cells per dish. If we plate 100 Petri dishes each with 1×10^6 *E. coli* cells in the presence of a normally lethal concentration of penicillin, and the mutation rate to penicillin resistance is 1×10^{-6}, we shall expect to find 37 dishes with no resistant mutants, 37 dishes with 1 resistant colony, 18 dishes with 2 resistant colonies, 6 dishes with three colonies, and 2 dishes with 4 colonies (or something close to this).

II. THE EQUATION OF GOLDIE AND COLDMAN

In the case of a potentially deadly infection (or tumor) we need to treat the disease before any resistant organisms arise, i.e., the probability of a cure is the probability that $\mu = 0$. This probability is given by

$$P_0 = \exp[-\alpha(N - 1)] \tag{7.3}$$

(Goldie and Coldman, 1979). Goldie and Coldman showed that the time taken to move from >95% probability of cure to <5% probability of cure was less than six doubling times. Thus, quite short delays in starting treatment can seriously compromise the effectiveness of chemotherapy.

III. MODELING ANTIBACTERIAL DRUG RESISTANCE

The ANTIBACT model was introduced in Chapter 6. It describes bacterial growth as exponential, with any required doubling time, and drug killing may follow any of ten dose–response relationships; it contains a simple description of the immune response, and includes three kinds of host toxicity. The ANTIBACT model allows for modeling of acquired resistance to two different drugs, including double mutants that are resistant to both drugs. The number of resistant cells at any particular time will be in agreement with the equation of Luria and Delbrück, and back-mutations are also modeled. Let us demonstrate the use of the ANTIBACT program to model drug resistance with a few examples.

A. Modeling a Bacterial Infection That Escapes the Immune Response

Let us consider the growth of a highly virulent bacterium that does not evoke an immune response. The initial inoculum is 100 cells, the doubling time (for wild type and for resistant mutants) is 2 h, and a lethal body burden is assumed in this case to be 3×10^9 cells. As shown in Table 7.1 and Figure 7.1, the pathogen replicates exponentially and results in death in less than 50 h. A high dose of trimethoprim administered at 24 h gives five log kill, and an increase in survival time of 67%. At the time of death of the host, the bulk of the pathogen is bact/S (wild type), as shown by Table 7.1. If the same high dose of trimethoprim is administered at 24 and 48 h, the life span is increased by 121% (Table 7.2), and most of the bacteria at the time of death are now trimethoprim-resistant (bact/R1). Table 7.2 (bottom) also makes the trivial point that if the trimethoprim is administered early, before the body burden exceeds five logs, and before resistant mutants have emerged, then a single dose of trimethoprim can cure this infection. Table 7.3 shows similar simulations with a different class of drug, the DNA gyrase inhibitor, ciprofloxacin, for which the mutation rate for resistance development is lower than for the antimetabolite, trimethoprim. With a low dose (Table 7.3, top) the sensitive organisms reach a lethal body burden at 64 h. For a higher dose (Table 7.3, bottom) death is again caused mainly by the wild type (bact/S) strain, though some bact/R2 (ciprofloxacin resistant), bact/R1 (trimethoprim resistant), and bact/R12 (doubly resistant) organisms are present. If ciprofloxacin is given daily for 3 d (Table 7.4, top; Figure 7.2), failure is now primarily caused by the ciprofloxacin-resistant bact/R2 organisms. Table 7.4 (bottom) again shows that a high enough dose of ciprofloxacin, used early enough, can eliminate this infection.

B. Modeling a Bacterial Infection That Causes an Immune Response

If we now assume that the bacterium is moderately immunogenic (setting the ANTIBACT immunogenicity parameter to 76.7), the small inoculum of 100 cells can now be eliminated without the aid of drugs (Table 7.5, top). However, when the inoculum is increased to 10,000 cells, despite the effects of the immune response, this is now a lethal infection (Table 7.5, middle). If we now treat this infection with a single dose of 50 mg/kg of ciprofloxacin, at 24 h, the drug eliminates about three logs of bacteria, and the rest are mopped up by the immune system (Table 7.5, middle). Contrast this with Table 7.3 (bottom), where in the absence of an immune response, 80 mg/kg of ciprofloxacin at 24 h was unable to eliminate a much smaller bacterial load.

C. Modeling a Two-Drug Combination

In this study, for the sake of simplicity, we again switch off the immune response by setting the ANTIBACT immunogenicity parameter to zero. The inoculum is set

Table 7.1

Antibacterial Model: Untreated Control					
Time	Bact/s	Bact/R1	Bact/R2	Bact/R12	CD8	Total
0	100	0	0	0	1	100
6	800	0	0	0	2	800
12	6,400	0	0	0	9	6,400
18	51,197	3	0	0	36	51,200
24	409,568	33	0	0	81	409,601
30	3,276,478	326	3	0	127	3,276,808
36	26,211,310	3,133	31	0	168	2,621,4474
42	209,686,368	29,239	292	0	205	209,715,904
48	1,677,400,064	267,324	2,673	0	237	1,677,699,968
49.75	3,076,400,128	508,147	5,081	1	245	3,076,900,096

Lethal bacterial load reached at 49.75 h

Increase in life span = 0%
Gross log kill = 0
Bacterial growth inhibition = −0.1%
Bacterial growth delay = 0 d

Antibacterial Model: Trimethoprim (1000 mg/kg) at 24 h					
Time	Bact/s	Bact/R1	Bact/R2	Bact/R12	CD8	Total
0	100	0	0	0	1	100
6	800	0	0	0	2	800
12	6,400	0	0	0	9	6,400
18	51,197	3	0	0	36	51,200
24	4	0	0	0	38	4
30	33	1	0	0	34	34
36	262	7	0	0	31	269
42	2,097	53	0	0	30	2,150
48	16,775	424	0	0	41	17,199
54	134,194	3,395	0	0	76	137,589
60	1,073,531	27,180	2	0	121	1,100,713
66	8,588,075	217,612	19	0	163	8,805,706
72	68,703,240	1,742,260	164	4	200	70,445,672
78	549,615,040	13,948,977	1,423	36	232	563,565,504
83	3,108,999,936	78,958,696	8,567	217	257	3,188,000,000

Lethal bacterial load reached at 83 h

Increase in life span = 67%
Gross log kill = 4.99
Bacterial growth inhibition = 100%
Bacterial growth delay = 1.4 d

to 10^5 bacteria. This infection causes death in under 30 h (Table 7.6, top). A maximum tolerated single dose (MTD) of ciprofloxacin (84 mg/kg) at 24 h prolongs survival, but fails to control the bacteria completely, and the wild type bacteria cause death at about 70 h (Table 7.6, bottom). If this MTD is administered every day for 5 d, the model predicted that the animal would succumb to the infection before the sixth

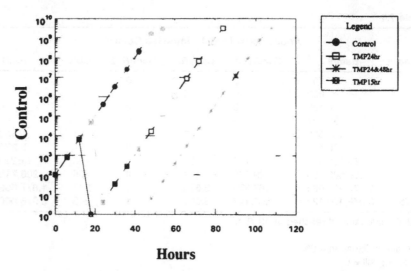

Figure 7.1 Effect of three different trimethoprim protocols on exponential growth of a bacterial population, as modeled by the ANTIBACT program.

dose could be administered, in this instance because of the selection of ciprofloxacin-resistant (bact/R2) organisms (Table 7.7). Table 7.8 shows the treatment of this same infection with maximum tolerated daily doses of trimethoprim; in this case again a single dose gave some survival prolongation, but failure resulted from proliferation of the wild type bacteria, and here again giving repeated daily doses resulted in selection of drug resistance (in this case bact/R1). Table 7.9 (top) illustrates the power of combination treatment with two non-cross-resistant drugs: a single treatment with trimethoprim and ciprofloxacin combined, both at their MTD, eliminated the infection within 24 h. The ability to combine the two agents, both at their MTD, assumes that they do not have overlapping toxicity; if toxicities do overlap, it is necessary to reduce doses. Table 7.9 shows what happens when the two drugs are given together, each at 50% of their MTD, daily for 9 d. The treatment fails because of the emergence of doubly mutant (bact/R12) cells, resistant to both drugs. However, adjusting the dose of trimethoprim upward very slightly resulted in a curative protocol (Table 7.10).

The recent resurgence of tuberculosis and other infectious diseases that were not long ago regarded as beaten has made it clear that the problem of drug resistance will be with us for a long time to come, perhaps indefinitely. Strategies for circumventing resistance will have to include the use of multiple drugs, preferably agents selected for mutual lack of cross-resistance. This raises questions of how best to combine the components of a combination protocol, especially when toxicities partially overlap, so that the drugs cannot be used together, simultaneously, at full doses. Is it best then to reduce doses, to use the compounds sequentially instead of simultaneously (so that dose reduction is unnecessary), or to employ an alternating protocol? As these questions become ever more complex, modeling will provide a valuable tool for unraveling the complexities of combination protocol design.

Table 7.2

	Antibacterial Model: Trimethoprim (1000 mg/kg) at 24 and 48 h					
Time	Bact/s	Bact/R1	Bact/R2	Bact/R12	CD8	Total
0	100	0	0	0	1	100
6	800	0	0	0	2	800
12	6,400	0	0	0	9	6,400
18	51,197	3	0	0	36	51,200
24	4	0	0	0	38	4
30	33	1	0	0	34	34
36	262	7	0	0	31	269
42	2,097	53	0	0	30	2,150
48	0	1	0	0	19	2
54	1	11	0	0	17	12
60	11	86	0	0	16	97
66	86	687	0	0	15	772
72	687	5,492	0	0	20	6,179
78	5,497	43,938	0	0	45	49,435
84	43,977	351,502	0	1	89	395,480
90	351,818	2,812,011	1	8	134	3,163,839
96	2,814,581	22,496,048	9	72	174	25,310,710
102	22,516,918	179,968,064	76	608	210	202,485,664
108	180,137,520	1,439,699,968	646	5,150	241	1,619,800,064
110	360,276,480	2,879,399,936	1,316	10,492	251	3,239,699,968

Lethal bacterial load reached at 110 h

Increase in life span = 121%
Gross log kill = 9.04
Bacterial growth inhibition = 100%
Bacterial growth delay = 2.5 d

	Antibacterial Model: Trimethoprim (1000 mg/kg) at 15 h					
Time	Bact/s	Bact/R1	Bact/R2	Bact/R12	CD8	Total
0	100	0	0	0	1	100
6	800	0	0	0	2	800
12	6400	0	0	0	9	6400
15	0	0	0	0	9	0

Bacteria eliminated at 15 h

IV. RESISTANCE TO ANTIVIRAL AGENTS

Antiviral chemotherapy presents essentially the same issues of drug resistance as antibacterial chemotherapy. In a sense the problem is more urgent with antivirals, because there are fewer good drugs, and they are generally more toxic. The ANTI-VIRL program, discussed in Chapter 6, models drug resistance similarly to ANTI-BACT. Figure 7.3 shows the example of a herpes simplex infection treated with a combination of low doses of acyclovir and vidarabine. At first the infection consisted almost entirely of the wild type virus/S, and the treatment gave a five-log reduction

Table 7.3

Antibacterial Model: Ciprofloxacin (30 mg/kg) at 24 h						
Time	Bact/s	Bact/R1	Bact/R2	Bact/R12	CD8	Total
0	100	0	0	0	1	100
6	800	0	0	0	2	800
12	6,400	0	0	0	9	6,400
18	51,197	3	0	0	36	51,200
24	2,934	0	0	0	69	2,934
30	23,470	2	1	0	79	23,473
36	187,757	22	4	0	113	187,783
42	1,502,025	209	33	0	154	1,502,267
48	12,015,963	1,915	267	0	192	12,018,145
54	96,125,832	17,234	2,157	0	225	96,145,224
60	768,991,488	153,188	17,407	3	255	769,162,112
64	3,075,899,904	653,596	70,037	15	273	3,076,600,064

Lethal bacterial load reached at 64 h

Increase in life span = 29%
Gross log kill = 2.14
Bacterial growth inhibition = 99.3%
Bacterial growth delay = 0.6 d

Antibacterial Model: Ciprofloxacin (80 mg/kg) at 24 h						
Time	Bact/s	Bact/R1	Bact/R2	Bact/R12	CD8	Total
0	100	0	0	0	1	100
6	800	0	0	0	2	800
12	6,400	0	0	0	9	6,400
18	51,197	3	0	0	36	51,200
24	1	0	0	0	53	1
30	6	0	0	0	47	6
36	50	0	0	0	42	50
42	400	0	2	0	37	402
48	3,198	1	17	0	37	3,216
54	25,587	5	133	0	52	25,724
60	204,691	41	1,061	0	90	205,793
66	1,637,499	359	8,487	2	134	1,646,347
72	13,099,739	3,132	67,900	16	174	13,170,786
78	104,795,848	27,139	543,207	140	210	105,366,336
84	838,350,272	233,814	4,345,737	1,208	241	842,931,008
87.75	3,075,000,064	895,913	15,940,418	4,631	259	3,091,800,064

Lethal bacterial load reached at 87.75 h

Increase in life span = 77%
Gross log kill = 5.72
Bacterial growth inhibition = 100%
Bacterial growth delay = 1.6 d

in viral load in about 2 weeks; however, by this time a small number of virus/R1 virions (acyclovir resistant) had appeared, and these proliferated to a life-threatening number. Note that in this simulation the drug-resistant virus replicates more slowly than the wild

Table 7.4

	Antibacterial Model: Ciprofloxacin (60 mg/kg) at 24, 48, and 72 h					
Time	Bact/s	Bact/R1	Bact/R2	Bact/R12	CD8	Total
0	100	0	0	0	1	100
6	800	0	0	0	2	800
12	6,400	0	0	0	9	6,400
18	51,197	3	0	0	36	51,200
24	21	0	0	0	59	21
30	168	0	0	0	52	168
36	1,345	0	1	0	48	1,346
42	10,759	2	6	0	53	10,767
48	4	0	2	0	59	6
54	35	0	15	0	52	50
60	283	0	118	0	47	401
66	2,261	0	946	0	45	3,208
72	1	0	281	0	43	282
78	7	0	2,250	1	41	2,258
84	59	0	18,000	5	51	18,065
90	475	0	144,000	43	86	144,518
96	3,802	1	1,151,978	366	129	1,156,147
102	30,412	10	9,215,646	3,113	170	9,249,181
108	243,290	87	73,723,704	26,376	206	73,993,456
114	1,946,292	737	589,777,984	222,753	238	591,947,776
118.75	10,096,003	3,981	3,059,300,096	1,203,740	261	307,059,9936

Lethal bacterial load reached at 118.75 h

Increase in life span = 139%
Gross log kill = 10.39
Bacterial growth inhibition = 100%
Bacterial growth delay = 2.9 d

	Antibacterial Model: Ciprofloxacin (84 mg/kg) at 24 h					
Time	Bact/s	Bact/R1	Bact/R2	Bact/R12	CD8	Total
0	100	0	0	0	1	100
6	800	0	0	0	2	800
12	6,400	0	0	0	9	6,400
18	51,197	3	0	0	36	51,200
24	0	0	0	0	52	0

Bacteria eliminated at 24 h

type. Toward the end of the simulation, virus/R2 (vidarabine resistant) and virus/R12 (doubly resistant) appeared, and virus/S (formed by reversion, or back-mutation, from resistant virions) reappeared (and grew more rapidly than the resistant virions).

V. MODELING DRUG RESISTANCE OF HIV

The phenomenon of drug resistance in RNA viruses (including retroviruses) appears to be about 100-fold more frequent than it is in bacteria, tumor cells, or

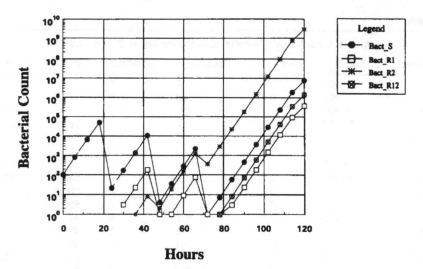

Figure 7.2 Emergence of drug resistance in a bacterial population treated with ciprofloxacin, as modeled by the ANTIBACT program.

DNA viruses, probably because RNA replication is inherently more error-prone than DNA replication. This means that a typical mutation rate at any particular locus will be about 1×10^{-4}. Since resistance to any given drug may be caused at multiple loci, the overall rate of development of resistance may be many times higher. Because the HIV genome consists of about 10^4 bases, and the mutation rate is one base in 10^4, on average each daughter virion will differ from the parent virion at one locus. Thus, instead of a coherent species of individuals that are genetically similar, HIV forms a "quasispecies" of related but highly variable individuals. At the beginning of this chapter, the question was raised whether, faced with this degree of genetic diversity, effective control of HIV with drugs for more than a few weeks is a realistic proposition at all. Because of the modest antiviral activity of the first-generation drugs, such as azidothymidine (AZT), there has been some controversy over whether these compounds prolong survival at all, although the balance of evidence seems to indicate that they do. This is a context in which modeling can provide predictions about how long a response it is reasonable to expect for particular pharmacokinetic and pharmacodynamic properties, and these predictions can help in making drug development decisions.

The VIROCHEM model was introduced in Chapter 6; it differs from ANTIVIRL in that it models the reciprocal destruction of virus and immune system, and it allows pharmacokinetics to be modeled explicitly. Figure 7.4 illustrates a sample simulation in which a patient was treated with low-dose AZT (300 mg, twice daily). At the start of the study the infection is assumed to have reached the antigenic diversity threshold, at which point it is essentially nonimmunogenic. According to the model, an untreated patient at this point survives for about 410 d before collapse of the immune system. The AZT treatment prolonged this period by a little over a year (430 d); at this point the CD4 count became very low, and the circulating viral load became very high. The circulating HIV was primarily AZT-resistant virus/R1, though drug sensitive virus was also still present.

Table 7.5

Antibacterial Model: Untreated Control for Immunogenic Infection						
Time	Bact/s	Bact/R1	Bact/R2	Bact/R12	CD8	Total
0	100	0	0	0	1	100
6	436	0	0	0	2	436
12	1,177	0	0	0	4	1,177
18	2,540	0	0	0	9	2,541
24	4,620	0	0	0	17	4,621
30	7,103	1	0	0	29	7,103
36	8,053	1	0	0	43	8,054
41.25	0	0	0	0	50	0

Bacteria eliminated at 41.25 h

Antibacterial Model: Control for High Inoculum						
Time	Bact/s	Bact/R1	Bact/R2	Bact/R12	CD8	Total
0	10,000	0	0	0	1	10,000
6	65,304	1	0	0	34	65,306
12	449,894	18	0	0	80	449,912
18	3,456,763	207	2	0	126	3,456,972
24	27,445,254	2,187	22	0	168	27,447,462
30	219,290,640	21,841	218	0	204	219,312,704
36	1,753,900,032	209,635	2,096	0	236	1,754,099,968
37.75	3,216,699,904	403,155	4,031	1	245	3,217,100,032

Lethal bacterial load reached at 37.75 h

Increase in life span = 4%
Gross log kill = 0.17
Bacterial growth inhibition = 33%
Bacterial growth delay = 0 d

Antibacterial Model: Ciprofloxacin (50 mg/kg) at 24 h						
Time	Bact/s	Bact/R1	Bact/R2	Bact/R12	CD8	Total
0	10,000	0	0	0	1	10,000
6	65,304	1	0	0	34	65,306
12	449,894	18	0	0	80	449,912
18	3,456,763	207	2	0	126	3,456,972
24	7,306	1	1	0	128	7,308
25	0	0	0	0	127	0

Bacteria eliminated at 25 h

Because of the successes of combination chemotherapy in treating bacterial infections and cancer, there is hope that a combination approach to AIDS may provide remissions of meaningful duration. Here again, the high mutation rate of HIV raises a question of whether this is a reasonable expectation. The VIROCHEM model was used to simulate treatment of a late-stage infection with a combination of AZT with the HIV protease inhibitor AG1343. When resistance developed to either agent, it was assumed for the purposes of these calculations to be 120-fold. AZT as a single agent (180 mg, three times daily) was predicted to give a response

Table 7.6

Antibacterial Model: Untreated Control for Nonimmunogenic Infection						
Time	Bact/s	Bact/R1	Bact/R2	Bact/R12	CD8	Total
0	100,000	0	0	0	1	100,000
6	799,984	16	0	0	54	800,000
12	6,399,742	255	3	0	103	6,399,999
18	51,196,900	3,059	31	0	147	51,199,992
24	409,566,944	32,634	326	0	186	409,599,904
29.75	3,004,499,968	296,756	2,967	0	219	3,004,800,000

Lethal bacterial load reached at 29.75 h

Increase in life span = 0%
Gross log kill = 0
Bacterial growth inhibition = -0.1%
Bacterial growth delay = 0 d

Antibacterial Model: Ciprofloxacin (84 mg/kg) at 24 h						
Time	Bact/s	Bact/R1	Bact/R2	Bact/R12	CD8	Total
0	100,000	0	0	0	1	100,000
6	799,984	16	0	0	54	800,000
12	6,399,742	255	3	0	103	6,399,999
18	51,196,900	3,059	31	0	147	51,199,992
24	404	0	3	0	119	408
30	3,234	0	26	0	108	3,260
36	25,869	3	208	0	115	26,081
42	206,951	29	1,664	0	145	208,644
48	1,655,579	264	13,308	2	182	1,669,153
54	13,244,369	2,375	106,467	19	217	13,353,230
60	105,952,856	21,106	851,741	169	247	106,825,872
66	847,606,272	185,735	6,813,962	1,487	274	854,607,424
69.75	3,108,900,096	719,981	24,993,802	5,767	289	3,134,700,032

Lethal bacterial load reached at 69.75 h

Increase in life span = 134%
Gross log kill = 6
Bacterial growth inhibition = 100%
Bacterial growth delay = 1.7 d

duration of 10.5 months. AG1343 alone (200 mg, three times daily) was predicted to give 15 months. The combination of AZT plus AG1343 (same doses) predicted a response duration of 49 months. This prediction was of interest for two reasons: first, the eventual viral breakthrough was of doubly resistant virus, and second, the predicted response duration was more than additive, since an additive response would have been 25 months. This favorable interaction of AZT with the HIV protease inhibitor was predicted to hold over a wide range of simulated doses. This particular simulation made the assumption that the single- and double-resistant HIV had similar virulence to the original drug sensitive strain. In fact, there is evidence that when the HIV protease gene mutates to give resistance to inhibitors, the mutant protease sometimes has decreased catalytic efficiency, which may translate into decreased

Table 7.7

	Antibacterial Model: MTD of Ciprofloxacin Daily for 5 d					
Time	Bact/s	Bact/R1	Bact/R2	Bact/R12	CD8	Total
0	100,000	0	0	0	1	100,000
6	799,984	16	0	0	54	800,000
12	6,399,742	255	3	0	103	6,399,999
18	51,196,900	3,059	31	0	147	51,199,992
24	404	0	3	0	119	408
30	3,234	0	26	0	108	3,260
36	25,869	3	208	0	115	26,081
42	206,951	29	1,664	0	145	208,644
48	2	0	133	0	116	134
54	13	0	1,060	0	104	1,073
60	105	0	8,480	2	100	8,586
66	837	0	67,837	15	120	68,689
72	0	0	5,403	1	100	5,404
78	0	0	43,225	11	113	43,236
84	0	0	345,789	96	148	345,886
90	4	0	2,766,262	824	186	2,767,090
96	0	0	220,330	70	140	220,401
102	0	0	1,762,608	595	178	1,763,204
108	1	0	14,100,586	5,045	213	14,105,631
114	7	0	112,802,440	42,604	244	112,845,048
120	0	0	8,984,616	3,572	172	8,988,188
126	1	0	71,875,504	30,011	208	71,905,520
132	23	0	574,992,576	251,540	239	575,244,160
137	184	0	3,252,499,968	1,476,909	263	3,254,000,128

Lethal bacterial load reached at 137 h

Increase in life span = 361%
Gross log kill = 16.11
Bacterial growth inhibition = 100%
Bacterial growth delay = 4.5 d

infectivity. As this kind of experimental data becomes available, it may readily be incorporated into the VIROCHEM model.

VI. PREDICTION OF RESISTANCE TO ANTICANCER DRUGS USING THE MODEL OF SKIPPER AND LLOYD

Skipper and Lloyd devoted many years to a detailed analysis of the large body of experimental data accumulated at the Southern Research Institute on treatment of experimental murine tumors with anticancer drugs. The general conclusions of this body of work have been summarized by Skipper (1986) and are also discussed in some detail in Jackson (1992). Lloyd related tumor cell kill during a course of treatment to an empirically determined parameter, K_s, the log cell kill of drug-sensitive cells per dose:

$$NLK = [K_s n] - [(\log 2)/DT \times (n - 1)I] \qquad (7.4)$$

Table 7.8

Antibacterial Model: Trimethoprim (1000 mg/kg) at 24 h						
Time	Bact/s	Bact/R1	Bact/R2	Bact/R12	CD8	Total
0	100,000	0	0	0	1	100,000
6	799,984	16	0	0	54	800,000
12	6,399,742	255	3	0	103	6,399,999
18	51,196,900	3,059	31	0	147	51,199,992
24	4,096	103	0	0	87	4,199
30	32,765	826	0	0	99	33,591
36	262,112	6,615	0	0	133	268,728
42	2,096,858	52,962	3	0	172	2,149,822
48	16,774,528	424,025	27	1	208	17,198,580
54	134,193,528	3,394,857	241	6	240	137,588,624
60	1,073,500,032	27,180,124	2,138	54	268	1,100,700,032
63	3,036,300,032	76,907,080	6,351	160	280	3,113,200,128

Lethal bacterial load reached at 63 h

Increase in life span = 112%
Gross log kill = 4.99
Bacterial growth inhibition = 100%
Bacterial growth delay = 1.4 d

Antibacterial Model: Trimethoprim (1000 mg/kg) Daily for 5 d						
Time	Bact/s	Bact/R1	Bact/R2	Bact/R12	CD8	Total
0	100,000	0	0	0	1	100,000
12	6,399,742	255	3	0	103	6,399,999
24	4,096	103	0	0	87	4,199
36	262,112	6,615	0	0	133	268,728
48	168	1,341	0	0	96	1,509
60	10,736	85,816	0	0	121	96,552
72	7	17,368	0	0	91	17,375
84	449	1,111,538	0	3	159	1,111,990
96	0	224,958	0	1	106	224,959
108	133	14,397,181	0	52	187	14,397,366
120	0	2,913,763	0	12	116	2,913,774
132	1,496	186,479,280	0	816	196	186,481,584
140.25	43,923	3,253,600,000	0	15,123	241	3,253,700,096

Lethal bacterial load reached at 140.25 h

Increase in life span = 372%
Gross log kill = 16.6
Bacterial growth inhibition = 100%
Bacterial growth delay = 4.6 d

where NLK = net log kill (which is gross log kill minus the amount of tumor regrowth during the treatment period), n = number of doses, DT = tumor doubling time, and I = interval between courses. Lloyd developed a computer program that incorporated this relationship with the Goldie and Coldman equations. This program allows simulations of the changing number and proportion of sensitive and drug-resistant

Table 7.9

Antibacterial Model: Trimethoprim (1000 mg/kg) and Cipro (84 mg/kg) at 24 and 48 h					
Time	Bact/s	Bact/R1	Bact/R2	Bact/R12	CD8	Total
0	100,000	0	0	0	1	100,000
12	6,399,742	255	3	0	103	6,399,999
24	0	0	0	0	56	0

Bacteria eliminated at 24 h

Antibacterial Model: Trimethoprim (500 mg/kg) and Cipro (42 mg/kg) Daily for 9 d					
Time	Bact/s	Bact/R1	Bact/R2	Bact/R12	CD8	Total
0	100,000	0	0	0	1	100,000
12	6,399,742	255	3	0	103	6,399,999
24	1,287	2	0	0	101	1,289
36	82,345	120	7	0	122	82,472
48	17	0	0	0	106	17
60	1,060	28	9	0	84	1,097
72	0	0	0	0	57	1
84	14	6	11	5	45	36
96	0	0	0	2	21	2
108	0	1	14	120	17	136
120	0	0	0	43	13	43
132	0	0	19	2,761	14	2,780
144	0	0	0	991	38	992
156	0	0	25	63,455	68	63,480
168	0	0	1	22,787	83	22,788
180	0	0	44	1,458,385	157	1,458,429
192	0	0	1	523,719	123	523,720
204	0	1	339	33,517,762	201	33,518,102
216	0	0	12	12,036,531	142	12,036,543
228	0	31	6,921	770,332,224	216	770,339,200
232	0	166	35,866	3,081,299,968	237	3,081,299,968

Lethal bacterial load reached at 232 h

Increase in life span = 680%

tumor cells during treatment with a single drug or a combination of two drugs that may be cross-resistant to varying extents. A version of this program, called SKIP-PER, is distributed on the computer disk supplied with the present author's earlier book (Jackson, 1992). Many of its conclusions are analogous to those we have already discussed in connection with anti-infective therapy.

1. Treatment with too-low doses results in rapid selection of resistance.
2. Cells selected for resistance to a particular drug often show cross-resistance to other drugs.
3. Pairs of non-cross-resistant drugs are often synergistic.
4. Disease may relapse from treatment with two drugs because of selection of double mutants.

Table 7.10

Antibacterial Model: Trimethoprim (600 mg/kg) and Cipro (42 mg/kg) daily for 3 d						
Time	Bact/s	Bact/R1	Bact/R2	Bact/R12	CD8	Total
0	100,000	0	0	0	1	100,000
6	799,984	16	0	0	54	800,000
12	6,399,742	255	3	0	103	6,399,999
18	51,196,900	3,059	31	0	147	51,199,992
24	407	1	0	0	94	408
30	3,255	8	0	0	87	3,264
36	26,040	67	2	0	96	26,109
42	208,314	537	17	0	129	208,868
48	2	0	0	0	85	2
54	13	1	0	0	75	14
60	106	9	1	0	66	116
66	848	70	7	1	59	925
72	0	0	0	0	30	0

Bacteria eliminated at 72 h

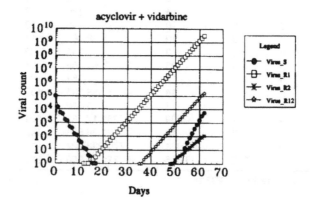

Figure 7.3   Drug resistance in Herpes simplex virus treated with combination chemotherapy. This simulation was run with the ANTIVIRL program.

5. Growth of resistant cells in absence of a continuing drug challenge (sometimes even despite presence of drug) may result in a significant number of drug sensitive revertants.

6. Drug-resistant cells may have doubling times that differ from that of the wild type (often longer).

The use of the SKIPPER program is illustrated with two examples. The first of these concerns drug resistance in relation to a cytokinetic parameter, cell loss factor. Goldie and Coldman (1983) pointed out that slowly growing solid tumors that have a high cell loss factor require more cell divisions to reach a given size than tumors in which cell loss is low. Thus, the slowly growing tumors are at greater risk of accumulating drug resistance mutations. This point is illustrated by the simulations

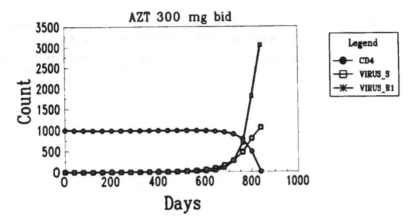

Figure 7.4 Emergence of drug resistance during treatment of an HIV infection with low-dose AZT, as modeled by the VIROCHEM program.

shown in Table 7.11. The upper part of the table shows the growth, as simulated by SKIPPER, of a murine tumor with a 12-h doubling time. No drug treatment was used, but the model was instructed to track resistance to 6-mercaptopurine (6-MP), selected because resistance to this drug is relatively frequent. After 18 doublings, the tumor has reached a mass of 2^{18} cells, i.e., something over 262,000, of which (in agreement with the equation of Luria and Delbrück) 980 are drug-resistant mutants. The lower portion of Table 7.11 shows a SKIPPER simulation of a murine tumor with a 12-h cell cycle time, but with a cell loss factor of 87.7% (meaning that on average, each cell cycle produces 1.123 daughter cells). This tumor has an effective doubling time of 72 h. After 108 cell cycle times, this tumor has reached the same cell mass as the tumor with no cell loss produced in 18 doublings. However, the slowly growing tumor took six times as many cell cycles to get there, and as a result it has acquired six times as many drug-resistant mutants. This illustrates the conclusion of Goldie and Coldman (1983) that slow-growing advanced clinical tumors have become very phenotypically heterogeneous.

We have not yet discussed the situation of collateral sensitivity, though examples of this phenomenon are encountered in anti-infective chemotherapy. This is the situation where, by mutating to resistance to one drug, a cell becomes more sensitive to a second drug (it is thus the opposite situation from cross-resistance). A simulation modeling an example of collateral sensitivity is summarized in Table 7.12. This considers the example of tumor cells that mutate to methotrexate (MTX) resistance by partial deletion of the reduced folate carrier protein, which is necessary for cellular uptake of MTX. Trimetrexate (TMTX), which, like MTX, is an inhibitor of dihydrofolate reductase (DHFR), enters cells by passive diffusion, so does not require the reduced folate carrier protein. Thus, cells that develop resistance to MTX by this mechanism are not cross-resistant to TMTX. However, because the reduced folate carrier is required for cellular uptake of natural folate cofactors as well, and since the higher the folate cofactor level in cells, the greater the degree of competition

Table 7.11

Time	Tumor/s	Tumor/R1	Tumor/R2	Tumor/R12	Tumor/Q	Total
			Program SKIPPER: Resistance to 6-MP (Alpha = 3E–4)			
0	1	0	0	0	0	1
12	2	0	0	0	0	2
24	4	0	0	0	0	4
36	8	0	0	0	0	8
48	16	0	0	0	0	16
60	32	0	0	0	0	32
72	64	0	0	0	0	64
84	128	0	0	0	0	128
96	256	0	0	0	0	256
108	511	1	0	0	0	512
120	1,022	2	0	0	0	1,024
132	2,043	5	0	0	0	2,048
144	4,086	10	0	0	0	4,096
156	8,170	22	0	0	0	8,192
168	16,336	48	0	0	0	16,384
180	32,666	102	0	0	0	32,768
192	65,318	218	0	0	0	65,536
204	130,610	462	0	0	0	131,072
216	261,165	979	0	0	0	262,144

Time	Tumor/s	Tumor/R1	Tumor/R2	Tumor/R12	Tumor/Q	Total
			Program SKIPPER: Cell Loss Factor = 87.8%			
0	1	0	0	0	0	1
72	2	0	0	0	0	2
144	4	0	0	0	0	4
216	8	0	0	0	0	8
288	16	0	0	0	0	16
360	32	0	0	0	0	32
432	64	0	0	0	0	64
504	127	1	0	0	0	128
576	253	3	0	0	0	256
648	506	6	0	0	0	512
720	1,011	13	0	0	0	1,024
792	2,020	28	0	0	0	2,048
864	4,035	61	0	0	0	4,095
936	8,059	132	0	0	0	8,191
1,008	16,098	283	0	0	0	16,381
1,080	32,156	607	0	0	0	32,763
1,152	64,232	1,293	0	0	0	65,525
1,224	128,302	2,746	0	0	0	131,048
1,296	256,282	5,810	0	0	0	262,093

with a DHFR inhibitor, folate cofactor transport-defective cells are actually more sensitive to TMTX.

In the simulation of Table 7.12 (top), the TMTX treatment at 96 h eliminates 3.7 logs of tumor/S cells, but merely maintains stasis of the R1 (trimetrexate-resistant) cells; the R2 (MTX-resistant) cells, of which there are only 2.2 logs at

Table 7.12

Program SKIPPER: TMTX (60 mg/kg) at 96 h and MTX (22 mg/kg) at 312 h						
Time	Tumor/s	Tumor/R1	Tumor/R2	Tumor/R12	Tumor/Q	Total
0	10,000	0	0	0	0	10,000
24	39,999	0	1	0	0	40,000
48	159,995	0	4	0	0	160,000
72	639,971	3	27	0	0	640,000
96	2,559,844	14	142	0	0	2,560,000
120	1,959	14	0	0	0	1,972
144	7,834	55	0	0	0	7,889
168	31,337	218	1	0	0	31,557
192	125,348	873	7	0	0	126,228
216	501,384	3,492	35	0	0	504,911
240	2,005,505	13,972	167	2	0	2,019,646
264	8,021,899	55,898	778	8	0	8,078,583
288	32,087,104	223,635	3,559	34	0	32,314,332
312	128,346,456	894,706	16,014	149	0	129,257,328
336	20,438	122,199	23,258	218	0	166,113
360	81,751	488,789	93,032	878	0	664,450
384	327,002	1,955,130	372,131	3,540	0	2,657,802
408	1,307,994	7,820,410	1,488,534	14,269	0	10,631,206
432	5,231,921	31,281,206	5,954,184	57,517	0	42,524,828

Program SKIPPER: MTX (22 mg/kg) at 96 h and TMTX (60 mg/kg) at 312 h						
Time	Tumor/s	Tumor/R1	Tumor/R2	Tumor/R12	Tumor/Q	Total
0	10,000	0	0	0	0	10,000
24	39,999	0	1	0	0	40,000
48	159,995	0	4	0	0	160,000
72	639,971	3	27	0	0	640,000
96	2,559,844	14	142	0	0	2,560,000
120	408	0	206	0	0	614
144	1,631	0	825	0	0	2,455
168	6,522	0	3,299	0	0	9,821
192	26,087	0	13,195	0	0	39,283
216	104,348	1	52,783	1	0	157,132
240	417,385	3	211,135	3	0	628,527
264	1,669,519	16	844,560	12	0	2,514,106
288	6,677,982	74	3,378,317	52	0	10,056,424
312	26,711,556	333	13,513,583	225	0	40,225,696
336	20,438	320	0	216	0	20,974
360	81,752	1,280	2	863	0	83,897
384	327,001	5121	14	3,454	0	335,589
408	1,307,985	20,486	73	13,814	0	1,342,358
432	5,231,860	81,951	363	55,258	0	5,369,432

this point, are completely eliminated. The MTX treatment at 312 h eliminates 4.4 logs of tumor/S cells, 0.44 log of R2 cells (which are tenfold resistant), and 1.46 logs of R1 cells — so the trimetrexate-resistant cells are threefold cross-resistant to MTX. By 432 h (18 d) there are 7.6 logs of tumor cells, of which the majority are R1 cells.

Table 7.12 (bottom) models the same drugs, at the same doses and treatment times, but the order of administration is reversed, with MTX now given first. The MTX gives the same degree of cell kill as before, eliminating all the R1 cells despite the threefold cross-resistance (because there are only 1.1 logs at 96 h). The most evident difference between the two protocols is seen at 312 h, where the TMTX eliminates over seven logs of methotrexate-resistant cells. As a result, by 432 h, there are eightfold fewer total tumor cells than for the previous simulation (the toxic effects on gut and bone marrow are identical). This pair of simulations demonstrates that quite small degrees of collateral sensitivity can have major effects on the efficacy of a combination protocol. It also shows the importance of drug sequencing and how sequence dependence can be influenced by the emergence of drug-resistant clones.

The methotrexate and trimetrexate combination is an example of one-way collateral sensitivity, i.e., MTX-resistant cells are collaterally sensitive to TMTX, but TMTX-resistant cells are cross-resistant to MTX. Collateral sensitivity always presents an opportunity for therapeutic benefit, and this is even more useful when the collateral sensitivity works both ways. Examples of this are known, for example, the combination of MTX with 6-MP. Cells may obtain their purine ribonucleotides by one of two metabolic pathways, the *de novo* pathway (which is inhibited by MTX) or the salvage pathway, utilizing the enzyme hypoxanthine-guanine phosphoribosyl-transferase (HGPRT), which is not inhibited by MTX. If cells become resistant to 6-MP by deletion of HGPRT (a common occurrence), they are now totally dependent for their purine biosynthesis on the MTX-sensitive *de novo* pathway. Some kinds of MTX-resistant cells have impaired uptake of the natural reduced folates, which are required for the *de novo* purine pathway; this forces the cell to be more dependent upon the salvage pathway, which is competitively inhibited by 6-MP. This combination was modeled by the SKIPPER program in Jackson (1992). The combination of MTX and 6-MP is used clinically for maintenance therapy of childhood acute leukemia. Clearly, this phenomenon of mutual collateral sensitivity is a highly desirable attribute of a combination protocol, and cytokinetic modeling provides a tool to explore how to get the most out of such combinations.

The examples of modeling with the SKIPPER program that are discussed above all involve simulation of the growth and treatment of murine tumors. Clearly, the simulation of human tumors can be performed in an analogous manner, though it is more difficult to obtain reliable cytokinetic parameter estimates. Birkhead et al. (1986) described simulation of clinical chemotherapy using a model very similar to that of Skipper; this study is discussed in Chapter 8.

VII. THE STOCHASTIC DRUG RESISTANCE MODEL OF COLDMAN AND GOLDIE

Because biological systems are so heterogeneous, some authors consider that it is overly simplistic to use strictly deterministic models to simulate tumor growth. In an actual tumor, whether in a mouse or a human, all the cells do not replicate with identical doubling times, the drug sensitivity of one cell will not be identical to that of its neighbors, and so on. The technique of stochastic modeling replaces fixed values for cytokinetic parameters with distributions. Coldman and Goldie (1986, 1987) based a stochastic model of tumor growth on an earlier model of MacKillop et al. (1983). At cell division, each stem cell may give rise to two cells, two differentiated cells, or one of each. Cells may be lost from any compartment. Unlike previous deterministic models, the rates of mutation to drug resistance were assumed to follow a β distribution. The authors wrote a computer program that incorporated equations describing rates of cell birth and death, rates of mutation to single- and double-drug-resistant mutants, and cell kill by drugs. This program was used to model treatment effects and to predict the effect of different treatment schedules on the numbers of surviving drug-resistant cells. One interesting prediction was that alternating schedules of administration of two drugs should give superior results to simultaneous or sequential administration.

VIII. USE OF THE EXCHEMO PROGRAM FOR MODELING DRUG RESISTANCE

Although the Skipper model provides a useful method for tracking the dynamics of drug-sensitive and -resistant cells during a course of chemotherapy, there are some problems where it is necessary to model the details of the cell cycle, which SKIPPER does not do. The EXCHEMO program, discussed in the previous chapter, is able to model problems that combine details of the cell cycle with development of drug resistance, and a few examples of its use are discussed. Since the SKIPPER program does not model cytokinetics and pharmacokinetics explicitly, it cannot predict schedule dependence. Thus, in SKIPPER, a dose of drug at 5 mg/kg twice daily will give the same gross log kill as a single 10 mg/kg dose, for both normal and malignant cells. However, this is not the case in real life, and the treatment schedule used may determine both the efficacy of the treatment and what cells, if any, survive the treatment. Figure 7.5 shows an EXCHEMO simulation of treatment of a rapidly proliferating murine tumor with carmustine (BCNU) and cytarabine (AraC). The BCNU is administered at time zero, and gives about 1.5 log kill of the sensitive cells and also of the cytarabine-resistant (R2) cells, but has very little effect on the R1 (BCNU-resistant) cells. At 24 h the AraC is administered as a single dose of 7 mg/kg. This gives over three log kill of the sensitive cells, completely eliminates the R1 cells, and has a lesser effect on the R2 cells. At the end of 96 h, 4.6×10^4 cells remain, of which 36% are AraC-resistant R2 cells. Figure 7.6 shows the system, treated with the same total doses of drug, except that the AraC dose is now divided into three portions of 2.33 mg/kg given at 24, 48, and 72 h. The total tumor burden at 96 h is now 1.2 logs lower than when the AraC was given as a single dose, and all the surviving cells are now R2 cells. This simulation makes the point that to predict the behavior of cell cycle-specific agents, we must use a model that includes details of the cell cycle.

Another run with the EXCHEMO model that emphasizes the importance of cytokinetic parameters in determining drug response is illustrated in Figures 7.7 and 7.8. This again models the effect of cytarabine on murine tumors. Figure 7.7 shows

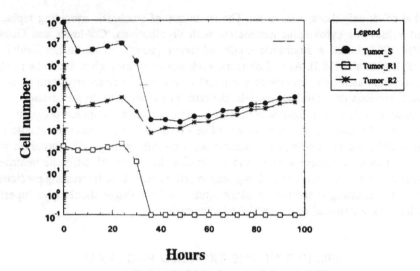

Figure 7.5   Use of the EXCHEMO program to model emergence of resistance to anticancer drugs. In this simulation R1 cells are resistant to carmustine (BCNU) and R2 cells are resistant to cytarabine (AraC).

the result for a tumor with 36-h cell cycle time, with no quiescent fraction, and a cell loss factor of zero. There is a net decrease in tumor burden of about 1.4 logs, and the AraC-resistant R1 cells decrease proportionately less. By contrast, the tumor of Figure 7.8 has a cell cycle time of 12 h, but a quiescent fraction of 55.5% (and cell loss factor again zero). Thus, the volume doubling time of this tumor is again 36 h. In this case, however, after the 45% of cells that were in cycle have been eliminated, the cell number almost plateaus, and the final tumor cell kill is much less than for Figure 7.7 (0.8 log). However, during the rapid kill phase, the small number of R1 cells is soon depleted, so that although the number of surviving cells is much higher than in Figure 7.7, they are all wild type cells. This emphasizes the important point that tumor cells that survive a course of treatment are not necessarily resistant cells; they may be potentially sensitive cells that have escaped for cytokinetic reasons. Thus, if we are to make realistic predictions about probable drug response, we must model cytokinetic "resistance" as well as genetic drug resistance.

IX. DRUG-RESISTANT CELLS AND BIOLOGICAL RESPONSE MODIFIERS

Biological response modifiers (BRMs) such as macrophage activators have the disadvantage that (in most cases) they are only able to eliminate quite small tumor burdens. Offsetting this disadvantage, to some extent, is that the activity of BRMs appears to be relatively little affected by many (though not all) of the common drug-resistance mechanisms. This has suggested the possibility that BRMs might be combined with chemotherapy in such a way that after conventional chemotherapy

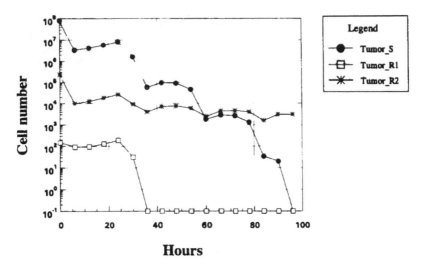

Figure 7.6 Combination chemotherapy of an experimental tumor with carmustine and cytarabine, showing the effect of dividing the cytarabine dose.

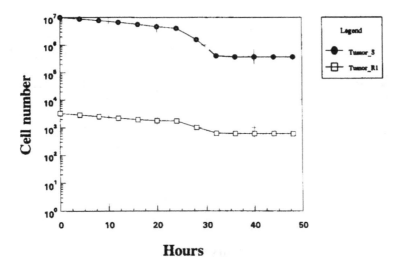

Figure 7.7 EXCHEMO simulation of cytarabine treatment of a tumor with long cell cycle time and zero quiescent fraction and cell loss factor.

has been used to eliminate the bulk of a tumor, an immunomodulator might help to mop up the drug-resistant cells that remain. Designing such studies raises many questions of how to select the optimal dosage and timing, and this is an area where cytokinetic modeling may help.

Figure 7.9 demonstrates the ability of the macrophage activator levamisole to eliminate cytarabine-sensitive and -resistant cells. Note that the original tumor

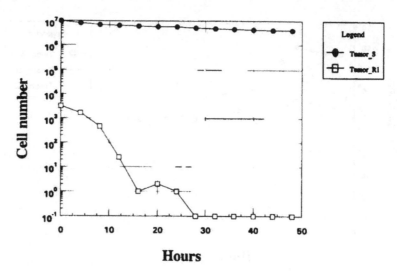

Figure 7.8 EXCHEMO simulation of cytarabine treatment of a tumor with short cell cycle time and high quiescent fraction and cell loss factor.

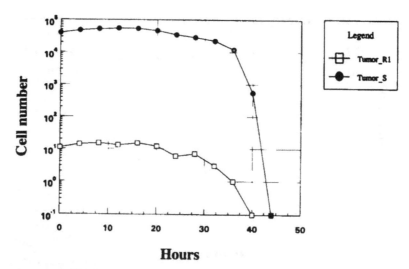

Figure 7.9 Elimination of drug-resistant cells by a macrophage activator, as modeled by the EXCHEMO program.

burden is small (4×10^4 cells), that cytarabine-resistant cells appear, but are effectively destroyed by the levamisole, and that resistance to the levamisole itself is not observed (i.e., there are no tumor/R2 cells). In experimental studies with mouse tumors, it is sometimes observed that tumor cells with acquired drug resistance may be more susceptible to immune destruction than the parental tumor cells. It is not known whether this is the case with human tumors.

X. DRUG-RESISTANT METASTATIC CELLS

Metastasis is such a universal feature of human cancer, and control of metastases is so essential to effective cancer treatment, that a useful cytokinetic model must include a description of metastasis. This means that if we want to study interactions of two drugs (an absolute minimum to do any useful modeling) we have to simulate eight tumor cell populations, i.e., tumor/S, tumor/R1, tumor/R2 and tumor/R12 (double-mutant) cells in the primary tumor, and the same tumor cell populations in the secondary tumors. The relationships of these eight populations are summarized in Figure 7.10. Note that tumor/S (wild type) cells may mutate to tumor/R1 cells (resistant to drug 1) and to tumor/R2 cells (resistant to drug 2). Tumor/R1 cells may have any desired degree of cross-resistance (or collateral sensitivity) to drug 2, and vice versa. Tumor/R1 and tumor/R2 cells may back-mutate (revert) to the drug sensitive phenotype, or they may undergo a second mutation to tumor/R12 cells, which are resistant to both drug 1 and drug 2. Since back-mutations from the tumor/R12 phenotype to tumor/R1 or tumor/R2 are likely to be rare events, they are not modeled. Any of the primary tumor cell populations may mutate to a metastatic phenotype, and the model considers the metastatic state as irreversible.

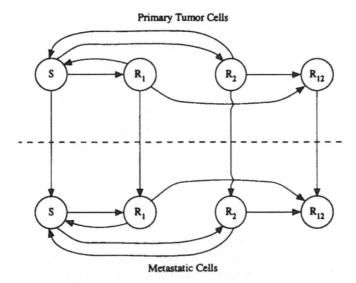

Figure 7.10 Interconversions of eight tumor cell populations. Sensitive cells may mutate to R1 or R2 cells, which are resistant to drugs 1 or 2, respectively, and these resistant cells may revert, by back-mutation, to the sensitive phenotype. R1 and R2 cells may undergo a second mutation to the R12 (double-mutant) cells, which are resistant to both drugs. Since this is a relatively rare event, back-mutation from the R12 phenotype is not modeled. Any of the four primary tumor cell populations may mutate to a metastatic phenotype, and the model considers the metastatic state as irreversible. (From Jackson, 1992. Reproduced with permission of Academic Press, Inc.)

From the clinical perspective, the significance of metastasis is that it makes a tumor more difficult to cure surgically. From a pharmacokinetic perspective, the significance of metastasis is that the metastatic tumor cells are exposed to a different concentration of drug than is the primary tumor. In the case of brain metastases, the difference in drug exposure may be considerable if the drug has poor penetration of the blood–brain barrier.

Table 7.13 shows the output from a PB-PK model of distribution of a 6-mg/kg i.v. bolus dose of Adriamycin in a mouse. This is a ten-compartment model, and for our present purposes the compartments of interest are the plasma, a subcutaneous solid tumor (which in this case is a small, well-vascularized tumor), the bone marrow and the GI mucosa (since these are the sites of host toxicity of the drug), and the brain (since we shall be modeling the growth of brain metastases). This output is from the PHYSIO program, which is available as part of the EXCHEMO package. Note that the Adriamycin distributes rapidly from plasma into most of the tissues. The tumor drug level greatly exceeds the plasma level, and the bone marrow drug level after 1 h is a little lower than the plasma level. The Adriamycin concentration in gastrointestinal mucosa at the later time points is lower still; gut cells contain a membrane pump, the P170 glycoprotein, that actively removes Adriamycin and many other xenobiotics from the cells. Note that the concentration of drug in the brain, at all time points, is lower than in most other tissues. Adriamycin is a hydrophilic drug that does not penetrate the blood–brain barrier well. In addition, the brain endothelial cells that comprise the blood–brain barrier are known to contain P170 glycoprotein, so it is possible that Adriamycin is actively pumped out of brain tissue.

The tissue distribution of carmustine (BCNU) after a dose of 8 mg/kg (i.v.) in a mouse is shown in Table 7.14. For this drug, note that brain concentrations are over 70% of plasma levels. Carmustine is a lipophilic drug that crosses the blood–brain barrier readily, presumably by passive diffusion. Note that carmustine also accumulates in adipose tissue, and has a prolonged gamma-phase half-life, perhaps because the fat soluble compound only slowly equilibrates back into the aqueous phase and becomes available for renal or hepatic elimination.

Figure 7.11 traces the effect of this two-drug combination on five tumor cell populations: tumor/S, tumor/R1 (Adriamycin resistant), tumor/R2 (carmustine resistant) in a primary tumor, and tumor/S and tumor/R1 secondaries. In this study there were no metastatic tumor/R2 cells present, and there were no tumor/R12 cells present in either the primary or secondary growths. The tumor rapidly metastasized to brain, and a few Adiamycin-resistant cells (designated as Met/R1 cells in Figure 7.11) appeared in brain. The Adriamycin treatment at time zero gave 1.4 log kill of the primary tumor/S cells, and 1.5 log kill of primary R2 (carmustine-resistant) cells, but only 0.14 log kill of the metastatic tumor/S cells. The carmustine treatment at 24 h killed 1.9 logs of primary tumor/S cells, and because carmustine crosses the blood brain barrier, it also killed 1.7 logs of metastatic tumor/S cells. The carmustine treatment also eliminated the R1 cells, both primary and metastatic.

This final simulation illustrates several points relevant to the science of modeling drug effects. The essence of this simulation was the question, how difficult is it going to be to eliminate metastatic tumor cells, including drug-resistant cells, in the

Table 7-13 PHYSIOPLEX: Physiological Pharmacokinetic Model Modeling Adriamycin in Human

6-mg/kg I.V. Bolus				
Time	Plasma	Muscle	Adipose	Tumor	Brain
1.0	658.3	3,902.1	1,754.8	9,431.3	1,343.5
4.0	172.8	5,500.6	1,748.2	10,474.7	1,777.4
8.0	119.4	5,044.6	1,225.4	7,865.2	1,536.5
12.0	90.1	4,162.7	888.5	5,820.8	1,219.9
16.0	69.8	3,329.4	671.8	4,411.5	952.7
20.0	54.7	2,635.4	520.1	3,408.1	743.8
24.0	43.1	2,080.5	407.4	2,664.1	582.7

Alpha-phase half-life = 0.10
Beta-phase half-life = 2.36
Gamma-phase half-life = 11.8

Time	Kidney	Bone marrow	Liver	GI mucosa	Gut lumen
1.0	151,296.0	20,136.9	23,778.6	27,435.2	19,110.9
4.0	98,844.6	8,058.8	5,958.2	7,247.5	23,230.4
8.0	60,607.0	5,107.7	4,051.6	4,911.2	15,145.3
12.0	44,098.7	3,805.1	3,042.3	3,680.6	10,830.3
16.0	33,809.9	2,937.5	2,354.3	2,845.9	8,202.0
20.0	26,387.2	2,298.4	1,844.0	2,228.3	6,366.5
24.0	20,756.3	1,810.1	1,453.0	1,755.5	4,995.7

Time	Lung	Spleen	Heart	Remainder
1.0	45,494.8	31,823.6	21,043.5	7,255.3
4.0	23,577.4	54,566.8	8,841.8	7,531.6
8.0	16,287.1	61,763.4	5,558.6	5,408.3
12.0	12,288.0	60,372.8	4,135.0	3,944.9
16.0	9,526.3	55,373.1	3,191.0	2,983.0
20.0	7,467.5	48,970.2	2,496.3	2,306.8
24.0	5,886.7	42,293.7	1,965.8	1,805.7

Drug eliminated in urine = 5.5%
Drug eliminated in feces = 49.2%
Drug metabolized = 24.2%
Renal clearance = 5.55 ml/h/g
Plasma AUC = 2.96 µMh

brain? This particular study required the ability to carry out cytokinetic modeling with a model of tumor growth that included the feature of metastasis. Since metastasis is one of the most characteristic features of malignancy, and the feature that probably most complicates cancer treatment, a model that did not include this feature would be too simplistic to be of much use. This study required the ability to model drug combinations; since most effective cancer chemotherapy involves combinations, this feature, too, is essential. Modeling drug resistance is likewise essential, because acquired resistance is one of the most frequent causes of treatment failure. Finally, the fact that the EXCHEMO cytokinetic model can use as its input drug concentrations that have been computed by a PB-PK model means that such features of drug distribution as the blood–brain barrier can be factored into the calculations.

Table 7-14   PHYSIOPLEX: Physiological Pharmacokinetic Model Modeling Carmustine in Human

8-mg/kg I.V. Bolus				
Time	Plasma	Muscle	Adipose	Tumor	Brain
1.0	78,015.8	5,609.5	85,277.7	67,735.9	55,188.5
4.0	55,574.9	4,002.4	112,808.3	50,080.8	40,043.6
8.0	50,662.8	3,648.7	102,833.8	45,653.4	36,503.9
12.0	46,203.4	3,327.5	93,779.8	41,634.9	33,290.7
16.0	42,138.4	3,034.8	85,528.8	37,971.8	30,361.8
20.0	38,428.8	2,767.6	77,999.3	34,629.0	27,688.9
24.0	35,045.0	2,523.9	71,131.2	31,579.8	25,250.8

Alpha-phase half-life = 0.39
Beta-phase half-life = 1.67
Gamma-phase half-life = 17.1

Time	Kidney	Bone marrow	Liver	GI mucosa	Gut lumen
1.0	46,400.3	31,247.6	5,637.8	7,854.8	1,453.1
4.0	32,860.9	22,234.1	4,001.1	5,558.6	3,829.1
8.0	29,956.4	20,268.8	3,647.5	5,067.5	4,038.8
12.0	27,319.5	18,484.7	3,326.4	4,621.4	3,757.1
16.0	24,916.0	16,858.4	3,033.8	4,214.8	3,436.5
20.0	22,722.5	15,374.3	2,766.7	3,843.8	3,135.3
24.0	20,721.7	14,020.5	2,523.1	3,505.3	2,859.4

Time	Lung	Spleen	Heart	Remainder
1.0	43,699.9	7,848.0	77,413.0	73,775.9
4.0	31,519.9	5,557.8	55,600.1	55,676.2
8.0	28,734.0	5,066.5	50,685.4	50,753.7
12.0	26,204.7	4,620.6	46,224.0	46,286.2
16.0	23,899.2	4,214.0	42,157.2	42,213.9
20.0	21,795.3	3,843.1	38,445.9	38,497.6
24.0	19,876.1	3,504.7	35,060.6	35,107.8

Drug eliminated in urine = 45.6%
Drug eliminated in feces = 22.9%
Drug metabolized = 0%
Renal clearance = 0.37 ml/h/g
Plasma AUC = 2.88 µM/h

A good general rule of computer modeling is that a model should be as simple as is consistent with retaining the essential controlling features of the system under study. However, the essence of biological systems is that many of their most interesting properties emerge at high levels of complexity, and to obtain insight into the effects of perturbing complex biological systems, with drugs or anything else, we must have a quantitative description that retains enough of the complexity of the parent system to be meaningful. A pharmacodynamic model that couples PB-PK models of drug distribution to a cytokinetic description of drug effects is starting to approach a level of complexity where meaningful therapeutic questions can be studied, while at the same time it is composed of elements that are experimentally verifiable.

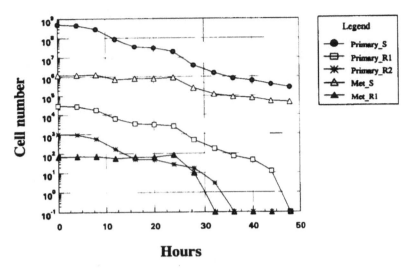

Figure 7.11 Treatment of a tumor that metastasizes to brain with a two-drug combination, modeled by the EXCHEMO program using input from a PB-PK model.

Figure

CHAPTER 8

Conclusions: The Virtual Clinical Trial

The preceeding chapters have emphasized the value of PB-PK modeling, of PK/PD modeling, of cytokinetic modeling, and of expert systems approaches as aids to preclinical drug development decisions. As the author hopes to demonstrate, the same techniques can serve as guides to clinical protocol design and evaluation also. In most cases, the techniques involved are not new; indeed, in some cases they go back more than 2 decades. However, there seems to be a widespread impression that these techniques are underutilized. The following recent quotes are typical of many:

> At present, only one relatively minor aspect of preclinical pharmacodynamics appears to be an established part of the drug development process. (Levy, 1993)

> Key opportunities exist for incorporation of PK/PD in drug development that could contribute to improving the efficiency of drug development. (Peck, 1993)

This underutilization of computational pharmacology techniques was probably because computer time used to be expensive, because people who make drug development decisions tended not to be computer literate, because the software involved was not very user friendly, and because biomathematicians and drug developers did not communicate effectively. These barriers are rapidly crumbling, and it is clear that future drug development will be much more computationally intensive. Ways in which modeling will contribute to clinical drug development decisions are illustrated with a small number of examples.

I. MODELING IN PHASE I CLINICAL TRIAL DESIGN

When a phase I clinical trial begins, what is known about the experimental drug has been obtained from animal studies, usually in rodents and dogs, and *in vitro* experiments, some of which may have been conducted with human material. What the phase I investigator has to establish is a suitable dose, route, and frequency of administration, what, if any, toxicity is observed in humans, and a number of classical

PK parameters. Some information on human metabolites is often obtained from a phase I clinical trial, as well. How can modeling contribute to this process? How can all the preclinical data be factored into clinical trial design?

Biomathematics can contribute to clinical trial design and analysis in two main ways: through statistics and through modeling. Statistical considerations, though crucial to well-designed and -conducted clinical trials, are beyond the scope of the present work, though a couple of computer programs deserve mention. The Clinical Trials Design Program, developed at the Johns Hopkins Oncology Center, is a commercially available, PC-based program that performs sample-size and power calculations for a variety of clinical study designs using established literature methods. Design-a-Trial (Wyatt, 1994) is an expert system to assist in the critique of experimental designs in clinical trials. However, our primary concern here is the contribution that can be made by modeling in enabling the clinical investigator to think through the consequences of different assumptions about dosage and scheduling.

A. Choosing a Starting Dose

One of the first decisions that has to be made when designing a phase I clinical trial is the choice of a starting dose. Safety is the major consideration, so typically a dose will be selected that is a generous submultiple of the lowest dose that showed any adverse effect in animals (e.g., for an anticancer drug, this may be 10% of the mouse MTD, assuming this dose [on a milligram-per-meter square basis] shows no toxicity in dogs [or sometimes 10% of the maximum "no effect" level in dogs, if this is lower]). Experience with hundreds of experimental drugs has shown that this is always safe (another drug development heuristic). However, while erring on the side of caution is always a popular choice, especially with regulatory authorities, starting a phase I trial at an unrealistically low dose may not always be the most appropriate, or even the most ethical, thing to do. For those drugs where phase I is conducted in patients (e.g., anticancer agents and AIDS drugs) we would like to minimize the number of patients exposed to subtherapeutic doses. From the point of view of the drug company, we would like to minimize the cost and duration of the trial by having as few escalation steps as possible between the starting dose and the dose recommended for phase II. The ideal starting dose, then, is the dose that is as close as possible to the probable human therapeutic dose that does not compromise safety. One way to use modeling to approach this ideal is to emphasize blood levels rather than doses; we know what blood level of drug (or what AUC) corresponds to a safe level, or a therapeutic level, in a mouse, so we can use interspecies scaling of the preclinical data to predict the appropriate human dose.

The starting dose for a phase I clinical trial may be schedule dependent. Collins et al. (1987) discuss the example of drugs where certain toxic effects are observed when the plasma drug concentration crosses a threshold value; for such drugs, a much greater AUC may be achieved by continuous infusion than by repeated bolus administration (Figure 8.1). Thus if the starting dose for a continuous infusion phase I were based upon the infusion MTD, rather than the bolus MTD (which would be the conventional choice), a much higher starting dose may be selected. Collins et al.

(1990) illustrate this principle with the example of deoxyspergualin (Figure 8.2), showing that by using the accelerated entry dose (based upon the infusion MTD), the number of escalations required was nine, while with the conventional entry dose 21 escalations would have been required, adding about an extra 2 years to the duration of the study.

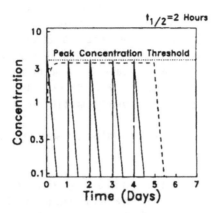

Figure 8.1 Role of a peak concentration threshold in determining the schedule dependence of a drug. In this example drug half-life is 2 h. Daily bolus injections (solid lines) produce peak concentrations just below the toxicity threshold (dotted line). Use of continuous infusion delivery (dashed line) enables the maximum tolerated concentration to be sustained for the duration of treatment, and the AUC for the infusion is eightfold higher than for the daily bolus schedule.

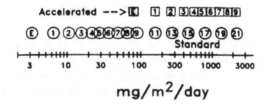

Figure 8.2 Phase I trial of deoxyspergualin started at an accelerated entry dose (E) of 80 mg/m²/d. The actual MTD was 2100 mg/m²/d, which could be reached in nine Fibonacci dose-escalation steps. If the standard entry dose (3.2 mg/m²/d) had been used, 21 escalation steps would have been required.

B. Dose Escalation

AUC values may also be used to guide escalation schemes. Again, the key reference is Collins et al. (1990). These authors point out that the ratio of AUC at MTD (human) to AUC at LD_{10} (mouse) is much less variable than the ratio of MTD (human) to LD_{10} (mouse); e.g., for brequinar the ratio of MTD to LD_{10} was 39, but the AUC (human to mouse) ratio was close to unity. If we adopt the hypothesis that AUC in humans at the MTD will be the same as AUC in mice at the LD_{10}, then the phase I doses may be rapidly escalated until the target AUC (or toxicity) is reached; Collins et al. recommend escalating in twofold steps, rather than according to the customary modified Fibonacci scheme (100% escalation, followed by 65, 50, and 40%, and then 30 to 35% for the remainder of the study). The experience with pharmacokinetically guided dose escalation has been summarized by Graham and Kaye (1993). Escalation based upon a pharmacodynamic end point would be even better, but this presupposes that a convenient biochemical or biophysical measure of drug effect is available. This is not always the case, while blood levels can nearly always be measured. For this reason, the development of pharmacodynamic tests is a high priority for clinical pharmacology research. Blood tests are acceptable, but noninvasive physical end points would be even better, e.g., positron-emission tomography (Tilsley et al., 1993) or NMR scans. NMR techniques in whole-body drug imaging have been mostly confined to ^{19}F NMR up to the present, but ^{13}C NMR and proton NMR are becoming more widely used (Maxwell, 1993).

C. Clinical Modeling Study 1: Optimizing the Protocol for an Antiviral Drug

Often, a drug is studied in phase I initially as a single dose, and when the peak plasma level, volume of distribution, clearance, plasma half-life, and other PK data have been obtained, the question arises, what is the optimum frequency of administration likely to be? Classical PK modeling is frequently used to study such questions. Figure 8.3 shows simulated plasma levels, obtained with PCNONLIN, based on human plasma PK of an antiviral drug. For this drug, the IC_{95} was estimated to be 500 ng/ml, and the aim of treatment was to ensure that the plasma drug concentration at the trough did not fall below this level. However, based upon the toxicokinetics, it was known that plasma levels above 1.5 μM trigger the onset of undesirable toxicity. As may be seen from Figure 8.3, a daily dose of 300 mg gave trough levels below what was required to block viral replication, even though peak levels were uncomfortably close to the toxicity threshold. It was clear that for a compound with these pharmacokinetic properties, no daily dose would achieve therapeutic trough levels without also giving toxic peak levels. Dividing the total daily dose into 150 mg twice daily keeps the peak plasma concentration at a safe level, but the trough levels still dip below the desired "floor" concentration of 500 nM (Figure 8.4). However, if we split the same total daily dose three ways and give 100 mg every 8 h, we can keep the trough plasma level above 500 nM without peak levels rising to dangerous spikes. As seen from Figure 8.5, it takes four doses to reach a steady state such that the amount of drug eliminated every 8 h is equal to the dose taken; before that, the plasma concentration does not always maintain a therapeutic level. If we are willing to let the total daily dose exceed 300 mg, it may be possible to find a twice-daily dose that will meet our needs. Figure 8.6 models 230 mg bid. Either this dose or 100 mg tid will keep the drug concentration within the "therapeutic window", but drug company marketing departments usually have a marked preference for a once-a-day formulation, or if that is not possible, twice a day, over more frequent dosage regimens, on the grounds of convenience and, it is believed, improved patient compliance (in other words, if treatment is once or twice a day, patients are less likely to miss a dose).

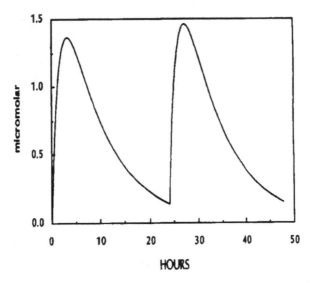

Figure 8.3 Simulation with PCNONLIN of plasma levels of an antiviral agent after daily administration of 300 mg. In this study the intent was to maintain minimal drug levels
above 500 ng/ml without exceeding a toxicity threshold of 1.5 µg/ml.

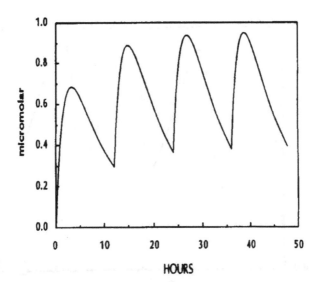

Figure 8.4 PCNONLIN simulation of the same antiviral drug studied in Figure 8.3, using a
dose of 150 mg every 12 h. Despite the divided dose, the trough levels still fall
below the desired minimum.

Regardless of whether the objective of modeling is to keep the drug level above
a particular minimum value, or below a toxic threshold (as in Figure 8.1), classical
PK modeling assumes that the plasma concentration is the relevant figure (or if not,
that the concentration at the effect site closely tracks the plasma level). A number

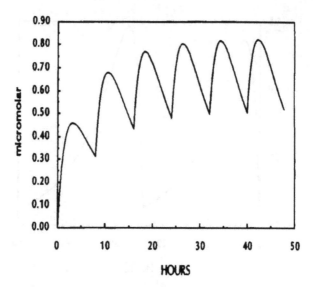

Figure 8.5 PCNONLIN simulation of the same antiviral drug studied in Figure 8.3. using a dose of 100 mg every 8 h. After the third treatment, plasma drug levels remain in the therapeutic range without exceeding the toxicity threshold.

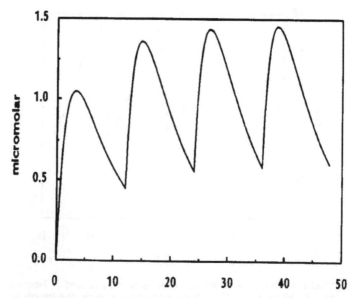

Figure 8.6. PCNONLIN simulation of the same antiviral drug studied in Figure 8.3. using a dose of 230 mg every 12 h. After the second dose, the plasma drug concentration remains in the safe and effective range. This schedule has the convenience of twice-daily treatment, though at the cost of using a higher total dose than for the three-times-daily regimen (Figure 8.5).

of PK fitting and modeling programs, in addition to PCNONLIN, can be used for this purpose, e.g., SIPHAR/win, MKMODEL, and MuPHARM (see Appendix). Obviously, if we need to know drug levels in some other compartment, a PB-PK model is necessary. Even though human drug levels are usually only available in plasma, a PB-PK model helps us to answer such questions as: based upon blood levels, are we delivering enough drug to the required site (e.g., tumor, skin, etc.)? Does the drug concentration stay at the required level for long enough to exert its pharmacological effect?

D. Clinical Modeling Study 2: A PB-PK Model

As an example of how a PB-PK model can contribute to clinical trial design, let us assume that we are developing an Adriamycin analogue, which, based upon the preclinical data, may be less cardiotoxic than Adriamycin. A PB-PK model has been created, based upon the preclinical data, and scaled into a human model, which has then been partially validated against preliminary phase I pharmacokinetics. The objective is to maximize exposure of the tumor to drug while minimizing the amount of exposure to drug of the heart. For drugs of this class, there is evidence that the antitumor activity correlates with AUC, but cardiotoxicity correlates with peak tissue level. Table 8.1 uses a PB-PK model to compare an i.v. bolus administration with the same dose given as a 3-h i.v. infusion. After an i.v. bolus, the predicted drug level in heart peaks early (Table 8.1, top). After a 3-h infusion, the drug level in heart peaked at 3.2 h, at 44% of the level that was predicted after the bolus dose. However, the predicted AUC values in tumor tissue were within 2% for the two administration schedules.

Mick and Ratain (1993) have proposed the use of a PK/PD model to accelerate the approach to MTD. The model they used to demonstrate the approach was based upon a data base of pharmacokinetic and pharmacodynamic data for etoposide. They claimed that their method could estimate the MTD with improved precision and with fewer patients treated at suboptimal doses.

E. Population Pharmacokinetics

One aspect of clinical pharmacology that we have not discussed previously is patient variability. This is not much of an issue with rodent studies, since the animals are often inbred, though this is not always the case when the experimental animals are dogs or monkeys. Population pharmacokinetics takes account of factors that may influence PK parameters — age, sex, race, weight, creatinine clearance, etc. The standard two-stage approach to defining a population model is to collect full data sets (usually requiring ten or more blood samples) from sufficient patients that an estimate can be obtained of the variance of the PK parameter estimates. In addition to being resource intensive, this approach tends to generate a high proportion of unusable data and may introduce bias into the final model, for reasons explained by Egorin (1992).

Table 8.1   PHYSIOPLEX: Physiological Pharmacokinetic Model Modeling X111815 in Human

1.03-mg/kg I.V. Bolus

Time	Plasma	Tumor	Bone marrow	GI mucosa	Heart
0	74,603	0	0	0	0
0.10	814	33,076	3,317	21,323	29,108
1.00	173	18,015	3,866	12,811	15,182
2.00	152	11,882	4,124	8,936	9,792
3.00	125	8,735	4,220	6,709	7,154
3.20	121	8,283	4,223	6,373	6,782
6.00	78	4,689	3,923	3,586	3,861
12.00	43	2,350	2,742	1,759	1,951
18.00	30	1,575	1,841	1,171	1,310
24.00	23	1,186	1,280	880	987

1.03-mg/kg 3-h I.V. Infusion

Time	Plasma	Tumor	Bone marrow	GI mucosa	Heart
0	0	0	0	0	0
0.10	47	107	11	68	94
1.00	244	5,496	764	3,654	4772
2.00	317	11,162	2,020	7,677	9,554
3.00	368	15,113	3,391	10,633	12,819
3.20	287	15,335	3,623	10,852	12,976
6.00	106	7,018	4,149	5,388	5,757
12.00	50	2,826	3,114	2,126	2,341
18.00	33	1,765	2,092	1,315	1,468
24.00	24	1,291	1,434	958	1,075

Note: Nanomolar concentrations.

A number of more sophisticated population pharmacokinetic modeling procedures have been developed. They can calculate maximum likelihood estimates of PK parameters that take into account our existing knowledge of these concomitant variables (Bayesian parameter estimation). Individualized dosing regimens can be devised by applying past experience with similar patients using a population pharmacokinetic model. An example of a program for carrying out calculations of this kind is P-PHARM (see Appendix). Perhaps the best known program for calculating population pharmacokinetic parameters is NONMEM (Beal and Sheiner, 1989). This program permits analysis of sparse data sets obtained in large numbers of patients to develop a description of the population pharmacokinetics. Another program suitable for population pharmacokinetic modeling is ADAPT II (D'Argenio and Schumitsky, 1979). A recent review (Egorin, 1992) complains that all software currently available for population PK modeling is computer intensive and user unfriendly.

If we know something about the spread of PK parameter values in our target population, and wish, for example, to calculate the minimum dose of our drug that will achieve therapeutic levels in 95% of the target group, population PK analysis is helpful. The other potential application of this approach is in the individualization of dosage for a particular patient:

Although the computing requirements are very demanding, Bayesian forecasting is an attractive means of adaptively modifying individual drug doses on the basis of a limited knowledge of a particular patient's pharmacokinetics, as set against a detailed database compiled from a much larger population... (Workman, 1993).

For a recent discussion of how to factor population pharmacokinetics into new drug development, see the article by Grasela and Antal (1993). Age-related changes in PK parameters are reviewed by Birnbaum (1991).

F. Sampling Theory

In clinical pharmacology studies, a serious practical limitation is the need to obtain a maximum of useful information from a limited number of patients' blood samples. Two general approaches to this problem have been described. The first is stepwise forward regression analysis, in which a detailed PK analysis is first performed on an initial group of patients (the training set). Analyses are performed for the AUC (the dependent variable) as a function of concentrations at each time point (the independent variable). The F-test is used to optimize the multiple regression. The final strategy is then optimized with a second full data set (the test set), and comparing estimated and measured values of AUC with this data set gives measures of bias and precision. The validated model can then give estimates of AUC from a limited number of samples, providing that the administration schedule is the same as was used to generate the model and that the samples were taken at time points that were used in the generation of the model. Ratain et al. (1988) discussed their Limited Sampling Model as applied to vinblastine and amonafide; in one of their studies, it was possible to estimate AUC from only two plasma samples. Kobayashi et al. (1993) have given a recent update of this approach.

Another approach, optimal sampling theory, has the advantage that it allows estimation of multiple PK parameters, not just AUC, and that it allows the use of samples collected at time points other than those used for the model generation (Peck and Perkins, 1984). This approach was reviewed by Egorin (1992), who states that its disadvantages are the very computer-intensive nature of the calculations, and the user-unfriendly software required.

II. MODELING IN PHASE II CLINICAL TRIAL DESIGN

Phase II studies are supposed to provide evidence of clinical efficacy: Does the compound work in humans? How active is the compound? For therapeutic agents (anti-infective or anticancer drugs), we also look for information about the spectrum of action. By this stage of clinical trials, the human PK data are known and the dose and frequency of treatment have been decided upon. Thus, where simulations of phase I clinical trials are usually concerned mainly with pharmacokinetics, phase II simulations generally involve pharmacodynamics. Some questions can be approached with quite simple models, but if we are to learn much from simulation of phase II clinical studies, we have to have a PK/PD model. Many of the models discussed in Chapters 5 to 7 may be applied just as easily to humans as to mice. However, it needs to be noted that these are deterministic models, so will not give any information about interpatient variability; we will be modeling a "typical" human subject, whatever that may mean.

We will consider two case studies, the first a simple PK/PD model where a single dose–response relationship is appended to a classical PK model, and the second a more complex example.

A. Clinical Modeling Study 3: Pharmacodynamics of an Antibiotic

Let us consider gentamycin, an antibiotic that is sometimes used to treat septicemia. In this case the target organ is the blood, so a PB-PK model or an effect-compartment model is unnecessary, and we can construct a PK/PD model that consists of a classical PK model with the addition of an equation that describes the dose–response curve for antibacterial activity of the antibiotic. The commercially available MKMODEL (see Appendix) includes a number of simple PK/PD models. In this example we assume a single-compartment PK model linked to a hyperbolic ("Emax") PD model. For this drug the minimum effective concentration is considered to be 2 μg/ml and the maximum safe concentration is 15 μg/ml. The example considered a 70-kg, 50-year-old subject with renal function of 73%. The volume of distribution was 18 l, and clearance was 4.7 l/h. The EC_{50} for the effect model was assumed to be 2 μg/ml. The gentamycin is administered as a 30-min infusion, every 8 h. Figure 8.7 shows MKMODEL's simulation: toward the end of each 8-h period, the gentamycin concentration drops below an effective antibacterial level. The antibacterial effect on this protocol varies between 40 and 90% of maximum. Although the PD model is rather rudimentary, MKMODEL makes it easy to study the effects of varying the dose, route, and schedule of administration, PK parameters, kidney function, etc. MKMODEL is also capable of slightly more complex PD modeling, e.g., effect-compartment models, modeling a delay factor, and replacing the simple Michaelis–Menten PD equation with the Hill equation.

B. Clinical Modeling Study 4: a Cytokinetic Simulation

Experimental tumors growing in mice tend to have cytokinetic parameters that differ significantly from autochthonous (formed where found) human tumors. Even xenografts of actual human cancers, when transplanted into immune-deprived mice, tend to speed up their growth rate, decrease their cell loss factor, and have a smaller quiescent fraction. This means that a drug whose activity is cell cycle dependent may often perform better against a transplanted tumor in a mouse than it will under actual clinical conditions. It is practically impossible to get a tumor to grow in a mouse with the same cytokinetics as it would grow in a human; even if it were possible, this would so prolong experimental chemotherapy studies that it would probably complicate anticancer drug testing as much as it would simplify it. In fact, there are a number of rat tumors whose cytokinetic behavior is reasonably close to that of an autochthonous human tumor, but they are not used much for experimental chemotherapy, probably for this reason. The solution seems to be to study our

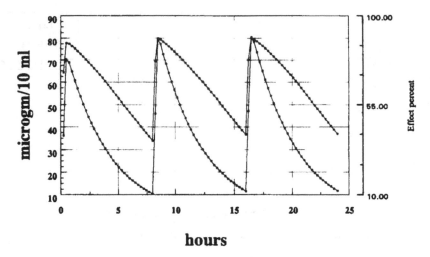

Figure 8.7   Use of MKMODEL to study pharmacokinetics (open circles) and pharmacodynamics (closed circles) of gentamycin. PK and PD parameter values are given in the text.

experimental drugs thoroughly against transplanted tumors, including xenografts, and then use a computational approach to predict how the experimental activity is likely to translate into clinical activity. Although cytokinetic modeling has had the capability of doing this for a number of years, surprisingly, this technique has been very little used in drug development.

As our fourth clinical case study, let us consider an anticancer drug that has demonstrated activity against a human colon carcinoma xenograft grown in nude mice. This experimental tumor has a cell cycle time of 24 h, a cell loss factor of 10%, and a quiescent (G_0) fraction of 5%. A highly malignant colon carcinoma growing in a human may have a cell cycle time of 48 h, a cell loss factor of 10% or more, and a G_0 fraction of 5% or greater; these last two parameters tend to increase with total tumor size, which can be simulated by the SKIPPER model. If our experimental drug gave a 90% (one log) kill per dose against the xenograft, what are we likely to see against the human tumor *in situ* (1) if the experimental drug is a cycle-nonspecific agent and (2) if it is a cell cycle-specific agent?

This calculation can be done with the EXCHEMO program (see Chapter 6). The appropriate cytokinetic parameters for the HCT-8 human colon carcinoma xenograft are entered, and a simulation is run with a dose of a phase-nonspecific agent selected to give a 1.0 gross log kill of tumor cells. Because the tumor has a volume doubling time of 30.2 h (based on a 24-h cell cycle time, and taking into account the 10% cell loss factor and the 5% quiescent fraction), the surviving tumor cells undergo some replication during the course of the experiment, and the net log kill at 72 h (i.e., allowing for tumor regrowth) is 0.29, i.e., tumor size after 72 h is 52% of the initial value. Now the cytokinetic parameters for the clinical tumor are entered and the simulation is repeated. This time the gross log tumor cell kill is only 0.7, but because the tumor volume doubling time is longer (74 h) the amount of regrowth is much less, so the net log kill at 72 h is 0.48, i.e., the final tumor size is 33% of the initial size. These results are summarized in Figure 8.8 (top).

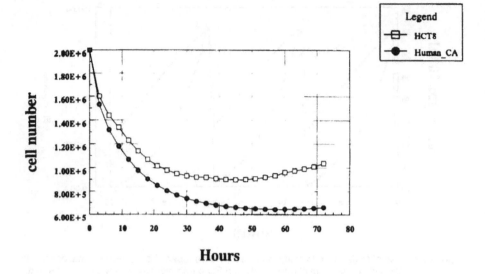

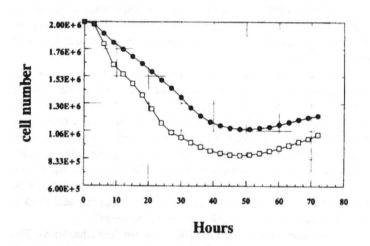

Figure 8.8 Cytokinetic simulation using the EXCHEMO program of effects of a cell cycle-nonspecific drug (top panel) and a cell cycle-specific agent (lower panel) on the HCT-8 human colon carcinoma grown as a xenograft in nude mice (open squares) and a human colon carcinoma (with different cytokinetic parameters) growing in a human patient.

The lower part of Figure 8.8 shows the result of a similar simulation for an agent that acts in G_2 phase of the cell cycle. Again, the dose was selected to give 1.0 log gross kill of the HCT-8 colon carcinoma xenograft; although the time-course differs, the overall effect over 72 h is similar to that given by the cycle-nonspecific drug (net log kill of 0.29). In this case the effect against the clinical tumor is less than against the experimental tumor: the size at 72 h is 60% of the starting value (0.22 log net kill). The lower regrowth rate of the clinical tumor has been more than counteracted by the lesser effect of the phase-specific drug against cells with a longer cycle time and with a higher quiescent fraction. This pair of simulations illustrates the point that a primary human tumor may be more or less sensitive than a xenograft growing in immunodeficient mice, although it is commonly assumed that exprimental tumor models have cytokinetic properties that tend to make them overpredict response.

For the sake of clarity, this simulation assumed that everything except the cytokinetic parameters was the same for the two tumors. In practice, the pharmacokinetics of the drugs will differ between mice and humans, and this will be an additional complication. However, the PK parameters can easily be changed in the simulation.

III. CLINICAL MODELING STUDY 5: PREDICTING TIME TO FAILURE

Responses to chemotherapeutic agents may be curative or subcurative. For diseases such as AIDS, where no curative treatment exists, treatment must continue until it is no longer effective, at which point it may be possible to elicit a further response by switching to another drug. Treatment may fail for a number of reasons, most often because of selection of a resistant organism, because of cumulative toxicity, or because of induction of metabolism of the drug.

Consider the colon carcinoma discussed in Example 4 that is growing with a 48-h cell cycle time, a loss factor starting at 10%, and an initial quiescent fraction of 5%. This tumor will be a highly aggressive one, with an initial volume doubling time of 74 h. However, both loss factor and quiescent fraction increase with tumor size, so that growth is Gomperzian. Let us assume that the tumor contains 10^8 cells at the time of diagnosis, and that 10^{12} cells will be lethal. This situation can be modeled using the SKIPPER program, which is based on the model of Skipper and Lloyd (Skipper, 1986). SKIPPER is included as part of the EXCHEMO program package (see Appendix). Figure 8.9 plots the uninhibited growth of this tumor, which reaches a lethal size 92 d after diagnosis.

Let us now assume that our experimental drug gives a two-log kill at its MTD and that the MTD can be administered every 21 d without cumulative toxicity. We know from *in vitro* studies that resistance occurs with a frequency of 10^{-7}, that the most common kind of resistant cell has a resistance factor of 20-fold, and that resistant cells have the same doubling time as the sensitive cells. How long will such a drug control the growth of this aggressive tumor before overgrowth of resistant cells causes treatment to fail? Using the SKIPPER model to simulate the effects of a maximum tolerated dose of this drug every 21 d (Figure 8.9) we see that the first three treatments give a good reduction in the number of tumor cells. Since a two-log kill corresponds to 6.6 doublings, and the effective doubling time is 74 h, so that 6.6 doublings takes a little over 20 d, if it were not for acquired drug resistance, this regimen ought to be able to maintain indefinite tumor stasis. As Figure 8.9 shows, the tumor is maintained in approximate stasis for about 2 months. However, the fourth treatment is less effective than the first three, and the fifth and subsequent treatments have almost no effect at all. The treated tumor reaches a lethal size at 156 d, corresponding to a 73% increase in life span.

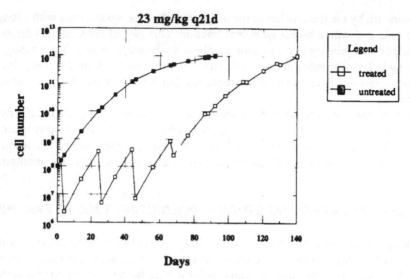

Figure 8.9 Use of the SKIPPER program to model treatment of a human colon carcinoma with maximum tolerated doses of a cycle-nonspecific agent every 21 d.

IV. MODELING IN PHASE III CLINICAL TRIAL DESIGN

The design of clinical combination chemotherapy studies involves so many variables that some clinical researchers may find that the simplifying assumptions required by a model make the exercise too simplistic to be helpful. The author believes that the more complex the system under study, the greater is the need for a modeling approach to isolate the variables, and think through the consequences of changing them, one at a time. Let us consider again the colon tumor that was modeled in Examples 4 and 5. Treatment with the experimental drug eventually failed because of the overgrowth of drug-resistant tumor cells. How helpful is it likely to be to combine the test drug with a non-cross-resistant agent? Figure 8.10 plots a SKIPPER simulation of combination chemotherapy using the test compound with a second drug that also gives about two logs tumor cell kill at its MTD. Both compounds are toxic to bone marrow, so it was necessary to reduce the dose of each agent to 50% of its MTD every 21 d. According to SKIPPER, this combination increased the survival from 90 d with no treatment and 156 d with either drug used as monotherapy (MTD every 21 d) to 206 d. Figure 8.10 shows the counts of the wild type tumor cells (tumor/s), cells resistant to drug 1 but still sensitive to drug 2 (tumor/R1), cells resistant to drug 2 (tumor/R2), and mutants resistant to both drugs (tumor/R12). At the beginning of the course of treatment, the tumor/s cells increase by about two logs every 21 d, and the treatment maintains approximate tumor stasis. The mutation rate for resistance to drug 2 is about 1.5 logs higher than for drug 1, so a few tumor/R2 cells are present from the outset. Tumor/R1 cells appear after a few days. Since both tumor/R1 cells and tumor/R2 cells are about one log resistant, each combination treatment gives about one log kill of these singly resistant cells.

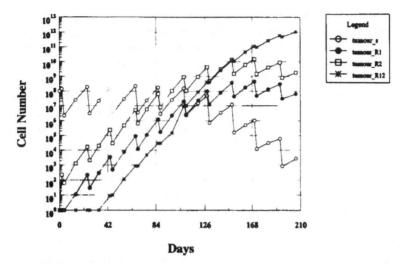

Figure 8.10   Use of the SKIPPER program to model treatment of a human tumor with a two-drug combination, showing eventual overgrowth by drug-resistant populations.

After about 6 weeks the first doubly resistant cells (tumor/R12) make their appearance. These are essentially unaffected by the treatment, and after 20 weeks they have become the dominant cell type in the tumor. At the time of death, they outnumber other cell types by almost three logs.

SKIPPER is capable of being used for many other kinds of simulation. It models drug effects on two normal tissues, bone marrow and intestinal epithelium (as well as keeping track of cumulative toxicity), so it is possible to ask how much additional therapeutic benefit might be obtained by combining two drugs whose organ toxicity targets were different.

Birkhead et al. (1986) described a model that appears to resemble in many respects the SKIPPER program discussed above, in that it incorporates the ideas of Skipper and of Goldie and Coldman. Birkhead's model assumes that each course of therapy kills a constant fraction of the sensitive cells, that tumor growth is exponential or Gompertzian, and that resistant cells may be present from the outset, or emerge by somatic mutation during the course of the treatment, or both. The authors discuss possible applications of their model. One application that they consider is to examine the consequences of adopting different treatment strategies, e.g., how does the effect of allowing a 30-d recovery period between cycles of treatment compare with a 50-d recovery period? The model was also used to explore the consequences of various degrees of cross-resistance between two drugs used in a combination protocol.

A. Analysis of Drug Combination Data

It is usually not possible to determine whether results of combined drug use in a clinical study are synergistic or antagonistic, in the rigorous sense of these words. It is, of course, possible to determine whether using drug 1 and drug 2 together is statistically better or worse than either drug alone, and these clinical effects are frequently described as therapeutic synergism or therapeutic antagonism. However, the fact that the effect of drug 1 plus drug 2 is superior to either drug singly and is therapeutically beneficial does not rule out the possibility that the interaction is merely additive or even subadditive. In preclinical studies, however, it is often feasible to do a rigorous determination of whether an interaction is additive, synergistic, or antagonistic. When a drug is ready to advance into Phase III clinical combination studies, the preclinical pharmacology department is often asked to determine the quantitative nature of the interaction between the test drug and drugs that it is likely to be combined with in phase III. Thus, while this is a nonclinical activity, it is one that takes place during clinical development, and this seems an appropriate place to discuss the quantitation of drug–drug interactions. The definition of what constitutes synergism depends on what constitutes additivity; the terminology is somewhat confused, and additivity may be defined in different ways (Greco et al., 1995). The most widely accepted definition is based upon the isobol concept: if dose D_1 of drug 1 gives a certain effect (e.g., 90% inhibition of viral growth) and dose D_2 of drug 2 gives the same effect, then the two drugs are additive if $(0.5 \times D_1) + (0.5 \times D_2)$ also gives 90% inhibition. If the combination gives a lower effect the drugs are antagonistic, and if the effect of the combination is greater than additive, the agents are synergistic.

A widely used method of quantifying drug interactions was described by Chou and Talalay (1984). This method expresses drug effects by a linearized version of the Hill equation known as the median effect equation:

$$\log(f_a/f_u) = m \log D - m \log D_m \tag{8.1}$$

where D is the dose, f_a and f_u are the fractions affected and unaffected, respectively, by this dose, D_m is the dose required to produce the median effect (i.e., the IC_{50}), and m is a Hill coefficient. If an experimental combination study is conducted with D_1 and D_2 at different concentrations (but keeping $D_1:D_2$ at a constant ratio), the data can be graphed to obtain m_1, $(D_m)_1$, m_2, $(D_m)_2$, $m_{1,2}$ and $(D_m)_{1,2}$, the last two parameters being for the mixture (Figure 8.11). For each effect level, the doses $(D_x)_1$, $(D_x)_{,2}$, and $(Dx)_{1,2}$ are calculated; these are concentrations of drug 1, drug 2, and the mixture, respectively, that inhibit by x%. Now, for each dose level of the mixture $(D_x)_{1,2}$, the contributions of drugs 1 and 2 are calculated, where, e.g., if the ratio of D_1 to D_2 is P/Q, then $(D)_1 = (D_x)_{1,2} \times P/(P + Q)$ and $(D)_2 = (D_x)_{1,2} \times Q/(P + Q)$. These results are then substituted into Equation 8.2 to calculate the combination index, CI:

$$CI = (D)_1/(D_x)_1 + (D)_2/(D_x)_2 + \alpha(D)_1(D)_2/(D_x)_1(D_x)_2 \tag{8.2}$$

If the effects of the two drugs are mutually exclusive (i.e., if D_1 and D_2 are known to act at the same binding site or if the median effect plots for D_1, D_2, and $D_{1,2}$ are parallel) then $\alpha = 0$ (so the last term of Equation 8.2 is dropped). If the effects of the two drugs are mutually nonexclusive (e.g., if they have independent binding sites) then $\alpha = 1$. CI is then plotted as a function of f_a, which gives an impression

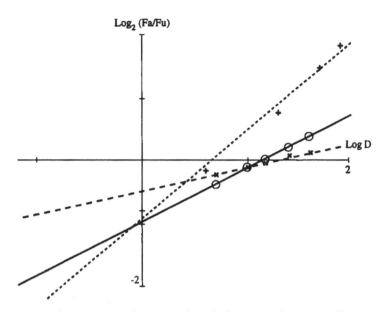

Figure 8.11   Median effect plots for a two-drug combination, analyzed by the method of Chou and Talalay (1982). X, 6-azauridine-5′-phosphate; O, PALA; +, combination.

of drug interaction at various levels of dose intensity (Figure 8.9). Note that CI = 1 corresponds to additivity, CI > 1 to antagonism, and CI < 1 to synergism. Although this procedure involves some tedious calculations, it can be easily automated using the program of Chou and Talalay, which is commercially available (see Appendix). Figure 8.11 shows data for the combination of the anticancer drugs PALA and 6-azauridine, which act at different points on the pyrimidine biosynthetic pathway. The median effect plots are not parallel, suggesting that the effects of the two drugs are mutually nonexclusive. The multiple drug-effect plot (Figure 8.12) shows that the PALA and azauridine combination is antagonistic for fractional inhibition levels of less than 0.3 and synergistic when fractional inhibition is greater than 0.3. At therapeutic levels (fractional inhibition > 0.9) the interaction is strongly synergistic.

Another technique that is becoming increasingly popular for quantitating drug–drug interactions is the response-surface approach; this involves fitting the experimental data to some equation that describes response in terms of concentrations of the two drugs and of an "interaction parameter" comparable to CI of Equation 8.2. Response-surface methods have been described by Carter and Wampler (1986), by Bunow and Weinstein (1990), and by Greco et al. (1990). These methods have a number of advantages: unlike isobol analysis, they do not require the use of a fixed end point; unlike the method of Chou and Talalay, they do not require that the two drugs be studied at a constant ratio; and they provide statistical measures of error of the parameter estimates, including the interaction parameter.

Bunow and Weinstein (1990) fit data to a version of the logistic equation, generalized to include an additional drug and one or more interaction parameters.

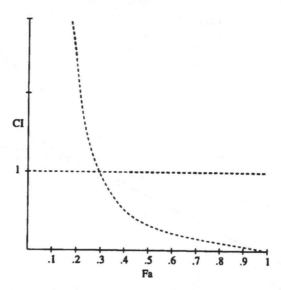

Figure 8.12 Multiple-drug effect plot for the combination of Figure 8.11, calculated by the method of Chou and Talalay (1982). The calculation assumes that the two agents are mutually nonexclusive.

They have described a computer program, COMBO, to fit experimental data; this program is designed to be used with the MLAB program package (see Appendix).

Greco et al. (1990) described a response-surface approach that can select from a variety of equations and weighting schemes. This method includes an interaction parameter, α; $\alpha = 0$ corresponds to additivity, $\alpha > 0$ implies synergism, and $\alpha < 1$ indicates antagonism. This technique (which also generates statistical confidence intervals, which allows an assessment of whether apparent synergism or antagonism is significant) has been automated by a computer program, SYNFIT (see Appendix). SYNFIT runs on a PC and produces graphical output, including isobols.

The increasing use of combination therapy, for cancer, AIDS, serious bacterial infections, arthritis, immunosuppression, hypertension, and many other conditions, has complicated the clinical development of those drugs that will be used in such combinations. Although it is probably unrealistic to expect detailed assessment of therapeutic drug interactions from clinical studies, it is important for optimal clinical use of such combinations to have rigorous preclinical data that quantitate the nature and extent of antagonistic or synergistic effects. This will probably entail more extensive computational analysis of drug combination data.

A final kind of drug interaction is that observed where one drug modulates the metabolism of the other. Antagonism may be observed, for example, if drug A induces enzymes that metabolize drug B. Synergism may be seen if drug A inhibits metabolism of drug B (e.g., by acting as a competing substrate for the same metabolic enzyme). Drug interactions of this type are readily recognized by the fact that blood levels of one drug will be affected by the presence of the other drug. This kind of drug interaction can thus be detected and analyzed by standard pharmacokinetic techniques.

V. EPILOGUE (SET JUST A FEW YEARS IN THE FUTURE)

As the sun sets over the Pacific Ocean, a drug designer comes in to work at Valhalla Pharmaceuticals, Inc., one of the biopharmaceutical companies strung along the coast of California. His current project is to design inhibitors of a enzyme called AICAR formyltransferase (AICARFT), since there is reason to believe that such compounds should have activity against rheumatoid arthritis. At the time of our story, the three-dimensional structures of a few percent of the 100,000 or so human enzymes have been solved and are available from the Protein Data Bank. For a significant fraction of the remainder, a complete or partial amino acid sequence is known and shows a relationship to a known structure, so that a plausible 3-D model can be constructed. The AICARFT structure is known, and before he left work this morning, our designer (let us call him Victor) left a docking program running on the Company's massively parallel supercomputer. This program scans a data base containing a few million chemical structures, and "docks" them into the AICARFT active site; that is, it examines each structure to see if it will fit into the binding site, allowing for the fact that each small molecule may have bonds that can be bent or rotated. Some of the small molecules in the data base will not fit the AICARFT active site at all; others can fit in multiple configurations (of which some may be energetically less favorable than others). For those compounds that fit, the program estimates the binding free energy. This requires computation of van der Waals forces, electrostatic interactions, hydrogen bond formation, internal energy of the bound ligand conformation, and the entropic factors involved in taking a small molecule out of aqueous solution and constraining it inside a rather lipophilic protein. At the time of our story, there is still some uncertainty in these calculations, so the binding score is calculated in several different ways. By the time Victor starts his evening's work, the supercomputer has produced a prioritized list of a few hundred compounds predicted to have the tightest binding. Now, AICARFT inhibitors have three main problems. First, they sometimes also bind to an enzyme called GARFT, which has the effect of weakening the biochemical consequences of AICARFT inhibition (because GARFT is an enzyme upstream in the same pathway, but we specifically want the substrate of AICARFT to accumulate in treated cells). Second, AICARFT inhibitors often inhibit another enzyme in a different pathway, DHFR, and this results in bone marrow suppression and other forms of toxicity. Third, AICARFT inhibitors may also inhibit thymidylate synthase (TS), which results in toxic effects similar to those caused by DHFR inhibitors. Fortunately, the 3-D structures of human GARFT, DHFR, and TS are also available, so it is possible to take the list of predicted AICARFT inhibitors and ask the computer for a list of the subset of compounds that do not have significant binding to GARFT, DHFR, or TS.

The resulting list of a few dozen compounds are structures in the chemical data base that have the right binding profile, but there is no guarantee that they are optimal structures. At this point, Victor runs another program known as a Monte-Carlo ligand generator. It examines the docking "hits" (predicted AICARFT inhibitors) and plays variations on these chemical themes — adds a methyl group at one position, takes off an amino from another position, and mixes and matches molecular fragments from the various hits. After each variation it recalculates the binding score, so that by the end of the run it has a list of about 50 compounds, each of which is predicted to be manyfold more potent as an AICARFT inhibitor than the original hits, and each of which is a novel, patentable compound previously unknown to chemistry.

Before the design phase is finished, there are three more jobs to do. The new structures are scored once again against the GARFT, DHFR, and TS structures, to make sure that the modifications have not given the new compounds the undesirable effect of binding tightly to these enzymes (some degree of binding to closely related enzymes is probably inevitable, and its effects will be factored in to the compound evaluation later). The inhibitor designs are run through the latest version of the REACCS program to see how synthetically feasible they are. Any compounds that cannot realistically be synthesized in a reasonable number of steps will be rejected; for the remainder, REACCS will suggest a synthesis. Finally, an expert system that predicts physical properties will examine the structures and predict values of log P, pK_a, and aqueous solubility. These properties will have a major influence on the pharmacology of the designed compounds. Any compounds that do not have a water solubility at pH 7 of at least 0.5 mM will usually be rejected at this stage.

Despite the power of the multiprocessor supercomputer, the drug design phase has taken several hours, and around midnight, when, according to legend, humans may turn into werewolves, Victor turns into a pharmacologist. About two dozen AICARFT inhibitor designs have survived the selection process thus far. They now have three further hurdles to cross: pharmacokinetics, metabolism, and pharmacodynamics. In the real-life laboratory at Valhalla Pharmaceuticals, after the finally selected compound design has been synthesized, many months of *in vitro* testing and animal pharmacology will be conducted. Victor's job is different: since the synthetic chemists and pharmacologists do not want to synthesize and study two dozen compounds, he must tell them which three or four compounds thay should concentrate on. An expert system for prediction of pharmcokinetics will examine the molecular weight, water and lipid solubility, electrostatic charge distribution at the pH of the stomach, the intestine, and of plasma, possible binding affinity to a number of active transport carriers, and likely binding to plasma protein fractions. Based on these properties, what kind of concentration will a reasonable daily dose produce in an inflamed human knee? Another expert system, perhaps the latest version of METABOLEXPERT, will examine the two dozen molecular designs and predict their most likely route and rate of metabolism in rats (likely to be the test species for the pharmacology studies) and humans. Now that we have an estimate of the concentration–time profile at the site of activity (e.g., the knee) and possible sites of toxicity (e.g., bone marrow), we are in a position to ask whether the amount of drug delivered will have the desired biochemical effect: do we inhibit AICARFT sufficiently to produce the desired elevation of its substrate, AICAR? Is the accumulated AICAR sufficient to have the required effect on A_2 receptors of neutrophils? Can this effect be produced without causing a major decrease in purine biosynthesis, which would depress RNA biosynthesis, and result in toxicity? Are any effects that the AICARFT inhibitor may have on GARFT, DHFR, or TS sufficiently minor that they will be unlikely to cause complications? All these pharmacodynamic questions can be studied using BioCHIMICA, using the concentration–time profiles in the appropriate compartments (joint and bone marrow) of a PB-PK model as input. If Victor were designing an anti-infective or anticancer drug he would also attempt to predict the likely incidence of drug resistance at this stage, but resistance to antiarthritic drugs is not usually observed.

Having passed the PK, metabolism, and PD hurdles, the candidate compounds must now enter toxicology. The COMPACT and TOPKAT tools available in our own time for predicting genotoxicity have, by the time of our story, been supplemented by a number of expert systems that can recognize chemical substructures likely to be associated with several kinds of organ toxicity. In addition to rule-based expert systems approaches, these programs use reasoning by analogy. A number of designs will be rejected at this stage on the grounds that they present an unacceptable risk of toxicity.

As the first rays of the rising sun appear over the mountains, the seven or eight surviving designs enter virtual clinical trials. Using a combined PB-PK/PD model, Victor asks what the probable dose and schedule of administration would need to be in humans, what the resulting concentration–time profile would be at the target site (phase I), what the resulting pharmacodynamic effect would be (phase II), how the predicted clinical response would compare with existing agents, and whether the new designs could be used favorably in combination with existing drugs (phase III).

Victor's working day ends with a review of his night's computations, culminating in his short list of three or four best designs for the AICARFT Project Team. This group, which includes medicinal chemists, pharmacologists, and clinical researchers, makes the final decisions on which compounds to synthesize and what the testing strategy should be.

Of course, this fable of a drug designer starting the night shift with a preliminary molecular design and finishing it with predicted clinical data is seriously unrealistic. What is interesting is the reason it is unrealistic. At the time of writing, software packages exist to do all the functions described in the fable. In some cases they do not do the job very well, but they are improving rapidly. What is actually unrealistic is the assumption that the process of design and development would be worth simulating from chemistry to clinic without experimental feedback. Even when the computational tools have been refined for a few more years, the uncertainties in their predictions will remain large, and if we string together the various techniques, uncertainties will accumulate until the signal gets lost in the noise. In real life, all these computational tools will be used, but there will be constant interaction between modeling predictions and experimental results. Even so, obviously not all compounds will work out in real life as the computer predicted. Currently, depending on the exact approach adopted, in a typical drug design and development program, the proportion of compounds found clinically active to total compounds synthesized ranges from one in a few hundred to one in several thousand. We do not have to increase that proportion by very much to pay for the cost of a supercomputer and a staff of computational pharmacologists.

A recent news report in *Science* discusses protein structure-based drug design, and quotes the early expectation that:

You need to see every atom in order to atomically tailor new drugs ... It's an information-based process, not a random process. It means we can get drugs faster to market and that they'll be better drugs.

The report continues:

So far, it has not worked out that way. There is no doubt now ... that the concept that just looking at a structure would be enough to design a new drug — is wrong. Few new drugs have reached the clinic by way of this high-tech approach, mostly because of obstacles long familiar to the pharmaceutical industry: solubility, efficacy, toxicity, and bioavailability ... (Hall, 1995)

What the article could have said, but did not, was that these long-familiar obstacles are at any rate no worse for structure-based designed drugs than for traditionally developed ones, and that despite these obstacles, an increasing number of rationally designed drugs are indeed reaching the clinic. However, the same article goes on to say.

The structural lab has become a permanent and prominent feature of the pharmaceutical industry.... There are now an estimated 50 to 60 structure labs in industrial settings...structural information can eliminate a lot of blind alleys in drug research.

In a similar way, computational pharmacology techniques (which are beginning to help us to surmount those long-familiar obstacles) will probably never allow us to make drug development so predictable that it becomes an engineering discipline. Used critically, in conjunction with experimental and clinical feedback, though, they can give us a much clearer picture of what our experimental drugs are doing, and should at any rate remove some of the guesswork from our exasperating and fascinating business.

References

Adolph EF (1949). Quantitative relations in the physiological constitutions of mammals. *Science* 109: 579-585.

Amidon GL (1983). Determination of intestinal wall permeabilities. In: *Animal Models for Oral Drug Delivery in Man:* In Situ *and* In Vivo *Approaches*, Crouthamel W and Sarapu C, Eds. Washington, D.C., American Pharmaceutical Association, 1-25.

Bachmann C and Colombo JP (1981). *Enzyme (Basel)* 26: 259-269.

Bachmann K (1989). Predicting toxicokinetic parameters in humans from toxicokinetic data acquired from three small mammalian species. *J. Appl. Toxicol.* 9: 331-338.

Bailar JC and Mosteller F (1992). *Medical Uses of Statistics, 2nd Ed.* Boston, NEJM Books.

Balant LP and Gex-Fabry M (1990). Review: physiological pharmacokinetic modeling. *Xenobiotica* 20: 1241-1257.

Bartel T and Holzhütter HG (1990). Mathematical modeling of the purine metabolism of the rat liver. *Biochim Biophys Acta* 1035: 331-339.

Beal SL and Sheiner LB (1989). *NONMEM User's Guides.* NONMEM Project Group, University of California, San Francisco.

Benowitz N, Forsyth RP, Melmon KL, and Rowland M (1974). Lidocaine disposition kinetics in monkey and man: II. Effects of haemorrhage and sympathomimetic drug administration. *Clin. Pharmacol. Ther.* 16: 99-109.

Birkett DJ, Mackenzie PI, Veronese ME, and Miners JO (1993). In vitro approaches can predict human drug metabolism. *Trends in Pharmacol. Sci.* 14: 292-294.

Birkhead BG, Gregory WM, Slevin ML, and Harvey VJ (1986). Evaluating and designing cancer chemotherapy treatment using mathematical models. *Eur. J. Cancer Clin. Oncol.* 22: 3-8.

Birnbaum LS (1991). Pharmacokinetic basis of age-related changes in sensitivity to toxicants. *Annu. Rev. Pharmacol. Toxicol.* 31: 101-128.

Bischoff KB (1986). Physiological pharmacokinetics. *Bull. Math. Biol.* 48: 309-322.

Bischoff KB and Brown RG (1966). Drug distribution in mammals. *Chem. Eng. Progr. Symp. Ser.* 66: 33-45.

Bischoff KB and Dedrick RL (1968). Thiopental pharmacokinetics. *J. Pharm. Sci.* 57: 1346-1351.

Bischoff KB, Dedrick RL, and Zaharko DS (1970). Preliminary model for methotrexate pharmacokinetics. *J. Pharm. Sci.* 59: 149-154.

Bischoff KB, Dedrick RL, Zaharko DS, and Longstreth JA (1971). Methotrexate pharmacokinetics. *J. Pharm. Sci.* 60: 1128-1133.

Bjorkman S, Wada DR, Stanski DR, and Ebling WF (1994). Comparative physiological pharmacokinetics of fentanyl and alfentanil in rats and humans based on parametric single-tissue models. *J. Pharmacokinet. Biopharm.* 22: 381-410.

Blumenson LE (1975). A comprehensive modeling procedure for the human granulopoietic system: detailed description and application to cancer chemotherapy. *Math Biosciences* 26: 217-239.

Bourne D (1995). *Mathematical Modeling of Pharmacokinetic Data.* 138 pp. Lancaster, PA, Technomic Publishing Co.

Boxenbaum H (1980). Interspecies variation in liver weight, hepatic blood flow and antipyrine intrinsic clearance in extrapolation of data to benzodiazepines and phenytoin. *J. Pharmacokinet. Biopharm.* 8: 165-176.

Boxenbaum H (1982). Interspecies scaling, allometry, physiological time, and the ground plan of pharmacokinetics. *J. Pharmacokinet. Biopharm.* 10: 201-226.

Boxenbaum H (1984). Interspecies pharmacokinetic scaling and the evolutionary-comparative paradigm. *Drug Metab. Rev.* 15: 1071-1121.

Boxenbaum H and D'Souza, R (1987). Interspecies pharmacokinetic scaling in reductionist and allometric paradigms. In: *Topics in Pharmaceutical Sciences,* Breimer DD and Speiser P, Eds. Amsterdam, Elsevier, 59-61.

Boxenbaum H and Ronfeld R (1983). Interspecies pharmacokinetic scaling and the Dedrick plots. *Am. J. Physiol.* 244: R768-R774.

Brown D and Rothery P (1993). *Models in Biology: Mathematics, Statistics, and Computing.* 688 pp. Chichester, John Wiley & Sons, Ltd.

Bungay PM, Dedrick RL, and Matthews HB (1981). Enteric transport of chlordecone (Kepone) in the rat. *J. Pharmacokinet. Biopharm.* 9: 309-.

Bunow B and Weinstein JN (1990). COMBO: a new approach to the analysis of drug combinations in vitro. *Ann. N.Y. Acad. Sci.* 616: 490-494.

Campbell DB and Ings RMJ (1988). New approaches to the use of pharmacokinetics in toxicology and drug development. *Human Toxicology* 7: 469-479.

Carter WH and Wampler GL (1986). Review of the application of response surface methodology in the combination therapy of cancer. *Cancer Treat. Rep.* 70: 133-140.

Cha S (1975). Tight-binding inhibitors -I. Kinetic behavior. *Biochem. Pharmacol.* 24: 2177-2185.

Chan KK, Cohen JL, Gross JF, Himmelstein KJ, Bateman JR, Tau-Lee Y, and Marlis S (1978). Prediction of Adriamycin disposition in cancer patients using a physiological pharmacokinetic model. *Cancer Treat. Rep.* 62: 1161-.

Chantret I, Barbat A, Dussaulx E, Brattain MG, and Zweibaum A (1988). Epithelial polarity, villin expression, and enterocytic differentiation of cultured human colon carcinoma cells: a survey of twenty cell lines. *Cancer Res.* 48: 1936-1942.

Chen HSG and Gross JF (1979). Physiologically based pharmacokinetic models for anti-cancer drugs. *Cancer Chemother. Pharmacol.* 2: 85-94.

Chou TC and Talalay P (1984). Quantitative analysis of dose-effect relationships: the combined effects of multiple drugs or enzyme inhibitors. *Adv. Enzyme. Regul.* 22: 27-55.

Coldman AJ and Goldie JH (1986). A stochastic model for the origin and treatment of tumors containing drug-resistant cells. *Bull. Math. Biol.* 48: 279-292.

Coldman AJ and Goldie JH (1987). Modeling resistance to cancer chemotherapeutic agents. In: *Cancer Modeling,* Thompson JR and Brown BW, Eds. New York, Marcel Dekker, Inc., 315-364

Collins JM, Leyland-Jones B, and Grieshaber CK (1987). Role of preclinical pharmacology in phase I clinical trials: consideration of schedule-dependence. In: *Concepts in Cancer Chemotherapy,* Muggia FM, Ed. Boston, Martinus Nijhoff, 129-140.

Collins JM, Grieshaber CK, and Chabner BA (1990). Pharmacologically guided phase I clinical trials based upon preclinical drug development. *J. Natl. Cancer Inst.* 82: 1321-1326.

Coltart DJ and Shand DG (1970). *Brit. Med. J.* 3: 731-734.

Conolly RB and Andersen ME (1991). Biologically based pharmacodynamic models. *Ann. Rev. Pharmacol. Toxicol.* 31: 503-523.

Cox DR and Snell EJ (1981). *Applied Statistics*. London: Chapman and Hall.

Cronin MT and Dearden JC (1989). Correlation of hydropobicity and molecular connectivities with toxicities of chlorobenzenes to different species. *Progress in Clinical and Biological Research* 291: 407-410.

D'Arginio DZ and Schumitsky A (1979). A program package for simulation and parameter estimation in pharmacokinetic systems. *Comput. Prog. Biomed.* 9: 115-134.

Darvas F (1987). METABOLEXPERT, an expert system for predicting metabolism of substances. In: *QSAR in Environmental Toxicology, Vol. II*. Keiser KLE, Ed. Dordrecht, Holland, Reidel Publishing Company.

Darvas F and Eker R (1990). Computer-assisted metabolite prediction, a help in development and drug design. In: *Pesticide Chemistry*, Freshe H, Ed. New York, VCH Publishing, 287.

Davies B and Morris, T (1993). Physiological parameters in laboratory animals and humans. *Pharmaceutical Res.* 10: 1093-1095.

Dayneka NL, Garg V, and Jusko WJ (1993). Comparison of four basic models of indirect pharmacodynamic responses. *J. Pharmacokinet. Biopharm.* 21: 457-478.

Dearden JC, Gregg CN, and Nicholson RM (1989). QSAR study of analgesic and anti-inflammatory potencies of commercially available non-steroidal anti-inflammatory drugs (NSAIDS). *Progress in Clinical and Biological Research* 291: 353-356.

Dedrick RL (1973). Animal scale-up. *J. Pharmacokinet. Biopharm.* 1: 435-461.

Dedrick RL, Bischoff KB, and Zaharko DZ (1970). Interspecies correlation of plasma concentration history of methotrexate. *Cancer Chemother. Rept.* (Part I) 54: 95

Dedrick RL, Forrester DD, and Ho DHW (1972). *In vitro-in vivo* correlation of drug metabolism-deamination of 1-β-D- arabinofuranosylcytosine. *Biochem. Pharmacol.* 21: 1-16.

Dedrick RL, Forrester DD, Cannon JN, Dareer SM, and Mellett LB (1973). Pharmacokinetics of 1-β-D-arabinofuranosylcytosine (Ara-C) deamination in several species. *Biochem. Pharmacol.* 22: 2405-2417.

Dedrick RL, Zaharko DS, and Lutz, RL (1973). Transport and binding of methotrexate *in vivo*. *J. Pharm. Sci.* 62: 882-890.

Dedrick RL, Zaharko DS, Bender RA, Bleyer WA, and Lutz RJ (1975). Pharmacokinetic considerations on resistance to anticancer drugs. *Cancer Chemother. Rep.* 59: 795-804.

Dedrick RL, Myers CE, Bungay PM, and DeVita VT (1978). Pharmacokinetic rationale for peritoneal drug administration in the treatment of ovarian cancer. *Cancer Treat. Rep.* 62: 1 -

DiMasi JA, Seibring MA, and Lasagna L (1994). New drug development in the United States from 1963 to 1992. *Clinical Pharmacology and Therapeutics* 55: 609-622.

Donaghey CE (1986). Cell kinetics simulation languages. *Bull. Math. Biol.* 48: 323-336.

Dressman JB, Amidon GL, and Fleischer D (1985). *J. Pharm. Sci.* 74: 588- 589.

Düchting W (1978). Control models of hemopoiesis. In: *Biomathematics and Cell Kinetics*, Valeron AJ and Macdonald PDM, Eds. Amsterdam, Elsevier/North Holland, 297-308.

Dullens HFJ, Van der Tol MWM, de Weger RA, and Den Otter W (1986). A survey of some formal models in tumor immunology. *Cancer Immunol. Immunother.* 23: 159-164.

Ebling WF, Wada DR, and Stanski DR (1994). From piecewise to full physiologic pharmacokinetic modeling: applied to thiopental disposition in the rat. *J. Pharmacokinet. Biopharm.* 22: 259- 291.

Egorin MJ (1992). Therapeutic drug monitoring and dose optimization in oncology. In: *New Approaches in Cancer Pharmacology: Drug Design and Development*, Workman P, Ed. Berlin, Springer-Verlag, 75-91.

Eigen M (1993). Viral quasispecies. *Scientific American* 269: 42- 49.

Everett NB, Simons B, and Lasher EP (1956). Distribution of blood (Fe^{59}) and plasma (I^{131}) volumes of rats determined by liquid nitrogen freezing. *Circ. Res.* 4: 419-424.

Farris FF, Dedrick RL, and King FG.(1988). Cisplatin pharmacokinetics: applications of a physiological model. *Toxicology Letters* 43: 117-137.

Fitzgerald GG (1993). Pharmacokinetics and drug metabolism in animal studies (ADME, protein binding, mass balance, animal models. In: *Integration of Pharmacokinetics, Pharmacodynamics, and Toxicokinetics in Rational Drug Development*, Yakobi A, Skelly JP, Shah VP, and Benet LZ, Eds. New York, Plenum Press, 23-31.

Franke R, Gruska A, and Presber W (1994). Combined factor and QSAR analysis for antibacterial and pharmacokinetic data from parallel biological measurements. *Pharmazie* 49: 600-605.

Frederick CB, Potter DW, Chang-Mateu MI, and Andersen ME (1992). A physiologically-based pharmacokinetic and pharmacodynamic model to describe the oral dosing of rats with ethyl acrylate and its implications for risk assessment. *Toxicology and Applied Pharmacology* 114: 246-260.

Furst DE (1993). Pharmacodynamic/pharmacokinetic relationships for rapidly acting drugs (NSAIDS) in rheumatoid arthritis: problems and preliminary solutions. In: *Integration of Pharmacokinetics, Pharmacodynamics, and Toxicokinetics in Rational Drug Development*, Yakobi A, Skelly JP, Shah VP, and Benet LZ, Eds. New York, Plenum Press, 193-200.

Gallo JM (1991). Pharmacokinetic models for anticancer and antiviral drugs following administration as novel drug delivery systems. In: *Advanced Methods of Pharmacokinetic and Pharmacodynamic Systems Analysis*, D'Argenio DZ, Ed. Plenum Press, New York, 21-36.

Garfinkel D and Hess B (1964). Metabolic control mechanisms. VII. A detailed computer model of the glycolytic pathway in ascites cells. *J. Biol. Chem.* 239: 971-983.

Gerlowski LE and Jain RK (1983). Physiologically-based pharmacokinetic modeling: principles and applications. *J. Pharm. Sci.* 72: 1103-1127.

Ghauri FY, Blackledge CA, Glen RC, Sweatman BC, Lindon JC, Beddell CR, Wilson ID, and Nicholson JK (1992). Quantitative structure-metabolism relationships for substituted benzoic acids in the rat. Computational chemistry, NMR spectroscopy and pattern recognition studies. *Biochem. Pharmacol.* 44: 1935-1946.

Gibaldi M and Perrier D (1982). Physiological pharmacokinetic models. In *Pharmacokinetics*, 2nd edn. New York, Marcel Dekker, 355-384.

Gibson GG and Skett P (1986). *Introduction to Drug Metabolism*, 293 pp. London, Chapman and Hall.

Gifford EM, Johnson MA, and Tsai CC (1994). Using the xenobiotic metabolism database to rank metabolic reaction occurrences. Molecular Connection 13: (1) 12-13; San Leandro, CA, MDL Information Systems, Inc.

Gilman AG, Mayer SE, and Melmon KL (1980). Pharmacodynamics: Mechanisms of drug action and the relationship between drug concentration and effect. In: *Goodman and Gilman's The Pharmacological Basis of Therapeutics*, 6th Edn. Gilman AG, Goodman LS, Gilman A, Eds. New York, Macmillan, 28-39.

Giovanella BC and Foght J (1985). The nude mouse in cancer research. *Advances in Cancer Research* 44: 69-120.

Goldie JH and Coldman AJ (1979). A mathematic model for relating the drug sensitivity of tumors to their spontaneous mutation rate. *Cancer Treat. Rep.* 63: 1727-1733.

Goldie JH and Coldman AJ (1983). Quantitative model for multiple levels of drug resistance in clinical tumors. *Cancer Treat. Rep.* 67: 923-931.

Goldstein GW and Betz AL (1986). The blood–brain barrier. *Scientific American* 255: 74-83.

Gonzalez FJ and Korzekwa KR (1995). Cytochromes P450 expression systems. *Ann. Rev. Pharmacol. Toxicol.* 35: 369-390.

Graham MA and Kaye SB (1993). New approaches in preclinical and clinical pharmacokinetics. In: *Pharmacokinetics and Cancer Chemotherapy*, Workman P and Graham MA, Eds. New York, Cold Spring Harbor Laboratory Press, 27-49.

Grasela TH and Antal EJ (1993). Pharmacoepidemiology, population pharmacokinetics and new drug development. In: *Integration of Pharmacokinetics, Pharmacodynamics, and Toxicokinetics in Rational Drug Development*, Yakobi A, Skell JP, Shah VP, and Benet LZ, Eds. New York, Plenum Press, 137-148.

Greco WR, Park HS, and Rustum YM (1990). Application of a new approach for the quantitation of drug synergism to the combination of cis-diamminedichloroplatinum and 1-β-Darabi nofuranosylcytosine. *Cancer Res.* 50: 5318-5327.

Greco WR, Bravo G, and Parsons JC (1995). The search for synergy: a critical review from a response surface perspective. *Pharmacol. Revs.* 47: 331-385.

Grevel J (1987). Kinetic-effect models and their applications. *Pharmaceutical. Res.* 4: 86-91.

Grieshaber CK (1991). Prediction of human toxicity of new antineoplastic drugs from studies in animals. In: The Toxicity of Anticancer Drugs, Powis G, Hacker MP, Eds. Oxford, Pergamon Press, 10-27.

Grindey GB, Moran RG, and Werkheiser WC (1975). Approaches to the rational combination of antimetabolites for cancer chemotherapy. In: *Drug Design*, Vol. 5, pp. 169-249, Ariens EJ, Ed. New York, Academic Press.

Guaitani A, Corada M, Lucas C, Lemoine A, Garattini S, Bartosek I (1991). Pharmacokinetics of fotemustine and BCNU in plasma, liver, and tumor tissue of rats bearing the Walker 256 carcinoma. *Cancer Chemother. Pharmac.* 28: 293-297.

Hall SS (1995) Industrial-strength protein structures. *Science* 267: 620-624.

Halpert JR (1995). Structural basis of selective cytochrome P450 inhibition. *Ann. Rev. Pharmacol. Toxicol.* 35: 29-53.

Hansch C and Leo A (1979). *Substituent Constants for Correlation Analysis in Chemistry and Biology.* Wiley, New York.

Hansch C, Kim D, Leo AJ, Novellino E, Silipo C, and Vittoria A (1989). Toward a quantitative comparative toxicology of organic compounds. *CRC. Crit. Rev. Toxicol.* 19: 185.

Harris PA and Gross JF (1975). Preliminary pharmacokinetic model for Adriamycin. *Cancer Chemother. Rep.* 54: 819-825.

Harvey RJ and Dev IK (1975). Regulation in the folate pathway of *Escherichia coli. Adv. Enzyme. Regul.* 13: 77-96.

Hawkins DR (1989). *Biotransformations*, Vol.2. Cambridge, Royal Society of Chemistry.

Heinemann V, Hertel LW, Grindey GB, and Plunkett W (1988). Comparison of the cellular pharmacokinetics and toxicity of 2′,2′-difluorodeoxycytidine and 1-β-D- arabinofuranosylcytosine. Cancer Res. 48: 4024-4031.

Heinmets F (1971). Computer simulation and analysis of tryptophan metabolism via kynurenine pathway in liver. *Comput Biol. Med.* 1: 323-336.

Herman RA and Veng-Pedersen P (1994). Quantitative structure- pharmacokinetic relationships for systemic drug distribution kinetics not confined to a congeneric series. *J. Pharm. Sci.* 83: 423-428.

Hewlett Parkard Co. (1974). *HP-45 Applications Book*, p.115. Cupertino, California, Hewlett Packard Co.

Hinderling PH, Schmidlin O, and Seydel JK (1984). Quantitative relationships between structure and pharmacokinetics of beta-adrenoceptor blocking agents in man. *J. Pharmacokinet. Biopharm.* 12: 263-287.

Hirono S, Nakagome I, Hirano H, Yoshii F, and Moriguchi I (1994). Non-congeneric structure-pharmcokinetic correlation studies using fuzzy adaptive least-squares: volume of distribution. *Biological and Pharmaceutical Bulletin* 17: 686-690.

Holford NHG (1990). *MKMODEL Version 4,* Manual. 236 pp. and software. Cambridge, Biosoft.

Holford NHG and Sheiner LB (1981). Understanding the dose-effect relationship. *Clin. Pharmacokin.* 6: 429-453.

Huang P, Chubb S, Hertel LW, Grindey GB, and Plunkett W (1991). Action of 2',2'-difluorodeoxycytidine on DNA synthesis. *Cancer Res.* 51: 6110-6117.

Ings RMJ (1990). Interspecies scaling and comparisons in drug development and toxicokinetics. *Xenobiotica* 20: 1201- 1231.

Jackson RC (1980). Kinetic simulation of anticancer drug interactions. *Int. J. Bio-Med. Comput.* 11: 197-224.

Jackson RC (1986). Kinetic simulation of anticancer drug effects on metabolic pathway fluxes: two case studies. *Bull. Math. Biol.* 48: 337-351.

Jackson RC (1989). The role of ribonucleotide reductase in regulation of the deoxyribonucleoside triphosphate pool composition: studies with a kinetic model. In: *Inhibitors of Ribonucleoside Diphosphate Reductase Activity.* Cory JG and Cory AH, Eds. pp. 127-150.

Jackson RC (1992). *The Theoretical Foundations of Cancer Chemotherapy Introduced by Computer Models.* 447 pp. and software. New York, Academic Press.

Jackson RC (1995). Toxicity prediction from metabolic pathway modeling. *Toxicology* 102: 197-205.

Jackson, RC (1996). Strategies for minimizing the development of HIV resistance to therapy: a modeling study. Submitted for publication.

Jackson RC and Harrap KR (1973). Studies with a mathematical model of folate metabolism. *Arch. Biochem. Biophys.* 158: 827-841.

Jenner P and Testa B (1980, 1981). *Concepts in Drug Metabolism, Parts A and B.* New York, Marcel Dekker.

Jusko WJ (1971). Pharmacodynamics of chemotherapeutic effects: dose-time response relationships for phase- nonspecific agents. *J. Pharm. Sci.* 60: 892-895.

Jusko WJ (1973). A pharmacodynamic model for cell cycle phase-specific chemotherapeutic agents. *J Pharmacokinet. Biopharm.* 1: 175-200.

Kaul S and Ritschel WA (1990). Quantitative structure- pharmacokinetic relationship of a series of sulfonamides in the rat. *Europ. J. Drug Metab. Pharmacokinet.* 15: 211-217.

Kawai R, Lemaire M, Steimer JL, Bruelisauer A, Niederberger W, and Rowland M (1994). Physiologically-based pharmacokinetic study on a cyclosporin derivative, SDZ IMM 125. *J. Pharmacokinet. Biopharm.* 22: 327-365.

King FG and Dedrick RL (1979). Pharmacokinetic model for 2-amino1,3,4-thiadiazole in mouse, dog and monkey. *Cancer Treat. Rep.* 63: 1939-1947.

King FG and Dedrick RL (1981). Physiologic model for the pharmacokinetics of 2'-deoxycoformycin in normal and leukemic mice. *J. Pharmacokinet. Biopharm.* 9: 519-534.

Klopman G (1992). MULTICASE. 1. A hierarchical computer automated structure evaluation program. *Quant. Struct. Act. Relat.* 11: 176-184.

Klopman G (1995). META: a new algorithm for the prediction of the products of biodegradation of chemicals. Abstract 137; Proceedings, Americal Chemical Society Meeting, Anaheim CA, 2-6 April 1995.

Knighton DR, Kan CC, Howland E, Janson CA, Hostomska Z, Welsh KM, and Matthews DA (1994). Structure of and kinetic channelling in bifuctional dihydrofolate reductase-thymidylate synthase. *Structural Biology* 1: 186-194.

Kobayashi K, Jodrell DI, and Ratain MJ (1993). Pharmacodynamic- pharmacokinetic relationships and therapeutic drug monitoring. In: *Pharmacokinetics and Cancer Chemotherapy*, Workman P and Graham MA, Eds. New York, Cold Spring Harbor Laboratory Press, 51-78.

Kobayashi Y, Kukuda Y, Nagatake H, Watanabe N, Wang X, and Ohno T (1995). Evaluation of *in vitro* toxicity of 12 MEIC compounds on human natural killer cell function. *In Vitro* Toxicology 8: 31-36.

Kremer et al. (1988). Pharm. Revs. 40: 1-47.

Krishan K and Andersen ME (1991). Interspecies scaling in pharmacokinetics. In: *New Trends in Pharmacokinetics*. Rescigno A and Thakur AK, Eds., New York, Plenum Press, 203-226.

Leal M, Yacobi A, and Batra VK (1993). Use of toxicokinetic principles in drug development: bridging preclinical and clinical studies. In: *Integration of Pharmacokinetics, Pharmacodynamics, and Toxicokinetics in Rational Drug Development*, Yakobi A, Skelly JP, Shah VP, and Benet LZ, Eds. New York, Plenum Press, 55-67.

Levy G (1993). The case for preclinical pharmacodynamics. In: *Integration of Pharmacokinetics, Pharmacodynamics, and Toxicokinetics in Rational Drug Development*, Yakobi A, Skelly JP, Shah VP, and Benet LZ, Eds. New York, Plenum Press, 7-13.

Lewis DFV (1992). Computer-assisted methods in the evaluation of chemical toxicity. *Rev. Comp. Chem.* 3:173-222.

Lewis DFV, Ioannides C, and Parke DV (1990). A retrospective study of the molecular toxicology of benoxaprofen. *Toxicology* 65: 33-

Lien EJ (1981). Structure-activity relationships and drug disposition. *Ann. Rev. Pharmacol. Toxicol.* 21: 31-61.

Lin JH, Sugiyama Y, Awazu S, and Hanano M (1982). *In vitro* and *in vivo* evaluation of the tissue-to-blood partition coefficient for physiological pharmacokinetic models. *J. Pharmacokinet. Biopharm.* 10: 637-647.

Lincoln T, Morrison P, Aroesty J, and Carter G (1976). Computer simulation of leukemia therapy: combined pharmacokinetics, intracellular enzyme kinetics, and cell kinetics of the treatment of L1210 leukemia by cytosine arabinoside. *Cancer Treat. Rep.* 60: 1723-1739.

Lipkowitz KB and Boyd DB, Eds. (1990, 1991, 1992). *Reviews in Computational Chemistry*, Vols. 1-3. New York, VCH Publishing.

Lloyd HH (1977). In: *Growth Kinetics and Biochemical Regulation of Normal and Malignant Cells*. Drewinko B and Humphrey RM, Eds. Baltimore, Williams and Wilkins, 455-469.

Luria SE and Delbrück M (1943). Mutations of bacteria from virus sensitivity to virus resistance. *Genetics* 28: 491-511.

Lutz RL, Dedrick RL, Straw JA, Hart MM, Klubes P, and Zaharko DS (1975). The kinetics of methotrexate distribution in spontaneous canine lymphosarcoma. *J. Pharmacokinet. Biopharm.* 3: 77-97.

Mackillop WJ, Ciampi A, Till JE, and Buick RN (1983). A stem cell model of human tumor growth: implications for tumor cell clonogenic assays. *J. Natl. Cancer Inst.* 70: 9-16.

Mariani G, Ferrante L, and Rescigno A (1991). Pitfalls in pharmacokinetic modeling of monoclonal antibody biodistribution in man. In: *New Trends in Pharmacokinetics*, Rescigno A and Thakur AK, Eds. New York, Plenum Press, 189-202.

Markin RS, Murray WJ, and Boxenbaum H (1988). Quantitative structure activity study on human pharmacokinetic parameters of benzodiazepines using the graph theoretical approach. *Pharmaceut. Res.* 5: 201-208.

Maxwell RJ (1993). New techniques in the pharmacokinetic analysis of cancer drugs. III nuclear magnetic resonance. In: *Pharmacokinetics and Cancer Chemotherapy*, Workman P and Graham MA, Eds. New York, Cold Spring Harbor Laboratory Press, 415-423.

Metzler CM and Weiner DL (1992). *PCNONLIN Version 4.0: Software for the Statistical Analysis of Nonlinear Models on Micros*. 380 pp. Lexington, Kentucky, SCI Software.

Mick R and Ratain MJ (1993). Model-guided determination of maximum tolerated dose in phase I clinical trials: evidence for increased precision. *J. Natl. Cancer Inst.* 85: 217-223.

Millenbaugh NJ, Kalns JE, Wientjes MG, and Au JLS (1995). Design of *in vitro* pharmacodynamic studies. *Proc. Amer. Assoc. Cancer Res.* 36: 361.

Mitsuhashi Y, Sugiyama Y, Ozawa S, Nitanai T, Sasahara K, Nakamura KI, Tanaka M, Nishimura T, Inaba M, and Kobayashi T (1990). Prediction of ACNU plasma concentration–time profiles in humans by animal scale-up. *Cancer Chemother. Pharmacol.* 27: 20-26.

Mordenti J (1986). Man vs. beast: pharmacokinetic scaling in mammals. *J. Pharm. Sci.* 75: 1028-1040.

Mordenti J and Green JD (1991). The role of pharmacokinetics and pharmacodynamics in the development of therapeutic proteins. In: *New Trends in Pharmacokinetics*, Rescigno A and Thakur AK, Eds. New York, Plenum Press, 411-424.

Morrison JF (1969). Kinetics of the reversible inhibition of enzyme catalyzed reactions by tight-binding inhibitors. *Biochim. Biophys. Acta*. 185: 269-286.

Morrison PF and Allegra CJ (1989). Folate cycle kinetics in human breast cancer cells. *J. Biol. Chem.* 264: 10552-10566.

Morrison PF, Lincoln TL, and Aroesty J (1975). Disposition of cytosine arabinoside and its metabolites: a pharmacokinetic simulation. *Cancer Chemother. Rep.* 59: 861-876.

Mulder GJ (1992). Glucuronidation and its role in regulation of biological activity of drugs. *Ann. Rev. Pharmacol. Toxicol.* 32: 25-49.

Neef C, Oosting R, and Meijer DK (1984). Structure-pharmacokinetics relationship of quaternary ammonium compounds. Elimination and distribution characteristics. *Naunyn Schmiedebergs Archives of Pharmacology* 328: 103-110.

Nicolini C, Milgram E, Kendall F, and Giaretti W (1977). Mathematical models for drug action and interaction *in vivo*. In: Growth Kinetics and Biochemical Regulation of Normal and Malignant Cells. Drewinko B and Humphrey RM, Eds. Baltimore, Williams and Wilkins, 411-433.

Nowack MA, Anderson RM, McLean AR, Wolfs TFW, Goudsmit J, and May RM (1991). Antigenic diversity thresholds and the development of AIDS. *Science* 254: 963-969.

Oh DM, Sinko PJ, and Amidon GL (1991). Predicting oral drug absorption in humans: a macroscopic mass-balance approach for passive and carrier-mediated compounds. In: *Advanced Methods of Pharmacokinetic and Pharmacodynamic Systems Analysis*, D'Argenio DZ, Ed. New York, Plenum Press, 3-11.

Ong S, Liu H, Qiu X, Bhat G, and Pidgeon C. (1995). Membrane partition coefficients chromatographically measured using immobilized artificial membrane surfaces. *Anal. Chem.* 67: 755-762.

Ozawa S, Sugiyama Y, Mitsuhashi J, and Inaba M (1989). Kinetic analysis of cell killing effect induced by cytosine arabinoside and cisplatin in relation to cell cycle phase specificity in human colon cancer and Chinese Hamster cells. *Cancer Res.* 49: 3823-3828.

Pang KS, Xu X, and St-Pierre MV (1992). Determinants of metabolite disposition. *Ann. Rev. Pharmacol. Toxicol.* 32: 623-669.

Parchment RE, Volpe DA, LoRusso PM, Erickson-Miller CL, Murphy MJ, and Grieshaber CK (1994). *In vivo-in vitro* correlation of myelotoxicity of 9-methoxypyrazoloacridine (NSC-366140, PD115934) to myeloid and erythroid hematopoietic progenitors from human, murine and canine marrow. *J. Natl. Cancer Inst.* 86: 273-280.

Pardridge WD (1988). Recent advances in blood–brain barrier transport. *Ann. Rev. Pharmacol. Toxicol.* 28: 25-39.

Peck CC (1993). Rationale for the effective use of pharmacokinetics and pharmacodynamics in early drug development. *Integration of Pharmacokinetics, Pharmacodynamics, and Toxicokinetics in Rational Drug Development*, Yakobi A, Skelly JP, Shah VP, and Benet LZ, Eds. New York, Plenum Press, 1-5.

Peck CC and Perkins SW (1984). Optimal sampling theory in a Bayesian context: a framework for choosing number and timing of clinical drug level measurements. *Clin. Pharmacol. Ther.* 35: 26.

Pfeifer S and Borchert HH, Eds. (1975-1983). *Biotransformation von Arzneimitteln. Vols. 1-5.* Berlin, VEB Verlag Volk und Gesundheit.

Pidgeon C, Ong S, Liu H, Qiu X, Pidgeon M, Dantzig AH, Munroe J, Hornback WJ, Kasher JS, Glunz L, and Szczerba T (1995). IAM chromatography: An *in vitro* screen for predicting drug membrane permeability. *J. Med. Chem.* 38: 590-594.

Pratt WB and Fekety R (1986). *The Antimicrobial Drugs.* 501 pp. Oxford, Oxford University Press.

Qu AS, Zheng X, Du CY, and An BS (1993). A cellular automaton model of cancerous growth. *J. Theoret. Biol.* 161: 1-12.

Ratain M, Staubus AE, Schilsky RL, and Malspeis L (1988). Limited sampling models for amonafide pharmacokinetics. *Cancer Res.* 48: 4127-4130.

Rey TD and Havranek WA (1992). Some aspects of using the SimuSolv program for pharmacokinetics applications. In: *Computer Simulation in Biology, Ecology and Medicine: 8th Prague Symposium.*

Riggs DS (1963). *The Mathematical Approach to Physiological Problems.* 445 pp. Baltimore, Williams and Wilkins Co.

Rogiers V, Sonck W, Shephard E, and Vercruysse A (1993). *Human Cells in In Vitro Pharmaco-Toxicology: Present Status within Europe.* VUB Press, Brussels.

Rowland M (1985). Physiologic pharmacokinetic models and interanimal species scaling. *Pharmacol. Ther.* 29: 49-68.

Rowles TK, Song X, and Ehrich M (1995). Identification of end points affected by exposure of human neuroblastoma cells to neurotoxicants at concentrations below those that affect cell viability. *In Vitro Toxicology* 8: 3-13.

Sacher GA (1959). Relationship of life span to brain weight and body weight in mammals. Ciba Foundation Colloquium on *Ageing* 5: 115-133.

Sanderson DM and Earnshaw DG (1991). Computer prediction of possible toxic action from chemical structure: the DEREK system. *Hum. Exp. Toxicol.* 10: 261.

Saunders L, Ingram D, and Jackson SHD (1989). *Human Drug Kinetics: A Course of Simulated Experiments.* 261 pp. Oxford, IRL Press.

Sawada Y, Hanano M, Sugiyama Y, Harashima H, and Iga T (1984). Prediction of volumes of distrribution of basic drugs in humans based on data from animals. *J. Pharmacokinet. Biopharm.* 12: 587-596.

Schellenberger W, Eschrich K, and Hofmann E (1981). Self- organization of a glycolytic reconstituted enzyme system. *Advin. Enzyme Regul.* 19: 257-284.

Scheuplein RJ, Shoaf SE, and Brown RN (1990). Role of pharmacokinetics in safety evaluation and regulatory considerations. *Ann. Rev. Pharmacol. Toxicol.* 30: 197-218.

Seelig A, Gottslich R, and Devant RM (1994). A method to determine the ability of drugs to diffuse through the blood–brain barrier. *Proc. Natl. Acad. Sci. U.S.A.* 91: 68-72.

Segel IH (1975). Enzyme Kinetics: *Behavior and Analysis of Rapid Equilibrium and Steady-State Enzyme Systems*, 957 pp., New York, John Wiley & Sons, Inc.

Segel LA (1984). *Modeling Dynamic Phenomena in Molecular and Cellular Biology.* 300 pp. Cambridge, Cambridge University Press.

Seither RL, Trent DF, Mikulecky DC, Rape TJ, and Goldman I D (1989). Folate pool interconversions and inhibition of biosynthetic processes after exposure of L1210 leukemia cells to antifolates. *J. Biol. Chem.* 264: 17016-17023.

Seydel JK and Schaper KJ (1981). Quantitative structure- pharmacokinetic relationships and drug design. *Pharmacol. Ther.* 15: 131-182.

Sheiner LB, Stanski DR, Vogel S, Miller RD, and Ham J (1979). Simultaneous modeling of pharmacokinetics and pharmacodynamics. *Clin. Pharmacol. Ther.* 25: 358-371.

Sigmund K (1993). *Games of Life: Explorations in Ecology, Evolution and Behaviour.* 244 pp. London, Penguin Books

Skipper HE (1986). On mathematical modeling of critical variables in cancer treatment. *Bull. Math. Biol.* 48: 253-278.

Skipper HE, Schabel FM, and Mellet LB et al. (1970). Implications of biochemical, cytokinetic, pharmacologic and toxicologic relationships in the design of optimal therapeutic schedules. *Cancer Chemother. Rep.* 54: 431-450.

Smith RL (1994). Predicting oral absorption rankings for new chemical entities using ideal solubility principles and *in vitro* pharmacology data. *Pharm. Res.* 11: S262.

Spector R, Spector AZ, and Snodgrass R (1977). Model for transport in the central nervous system. *Am. J. Physiol.* 232: R73-R79.

Struck MM (1994). Biopharmaceutical R&D success rates and development times. *Bio/Technology* 12: 674-677.

Sugita O, Sawada Y, Sugiyama Y, Iga T, and Hanano M (1982). Physiologically-based pharmacokinetics of drug–drug interaction: a study of tolbutamide-sulfonamide interactions in rats. *J. Pharmacokinet. Biopharm.* 10: 297-.

Swan GW (1987). Tumor growth models and cancer chemotherapy. In: *Cancer Modeling.* Thompson, JR and Brown BW, Eds. New York: Marcel Dekker, 91-179.

Tallarida RJ and Murray RB (1987). *Manual of Pharmacologic Calculations with Computer Programs*, 2nd edn. 297 pp. and software. New York, Springer-Verlag.

Talley G (1993). Rating health hazards at Los Alamos National Laboratory. *HDI Toxicology Newsletter*, No. 16, p.4. Rochester, NY, Health Designs, Inc.

Testa B and Jenner P (1976). *Drug Metabolism: Chemical and Biochemical Aspects.* New York, Marcel Dekker.

Testa B and Salvesen B (1980). Quantitative structure-activity relationships in drug metabolism and disposition: pharmacokinetics of N-substituted amphetamines in humans. *J. Pharm. Sci.* 69: 497-501.

Tew KD, Houghton PJ, and Houghton JA (1993). *Preclinical and Clinical Modulation of Anticancer Drugs.* 364 pp. Boca Raton, FL: CRC Press, Inc.

Thakur AK (1991). Model: mechanistic vs. empirical. In: *New Trends in Pharmacokinetics.* Rescigno A and Thakur A.K., Eds. New York, Plenum Press, 41-51.

Thorburn DR and Kuchel PW (1985). *Eur. J. Biochem.* 150: 371-386. Tichy M (1985). *QSAR in Toxicology and Xenobiochemistry.* 474 pp. Elsevier, Amsterdam.

Tilsley DWO, Harte RJA, Jones T, Brady F, Luthra SK, Brown G, and Price PM (1993). New techniques in the pharmacokinetic analysis of cancer drugs. IV positron emission tomography. In: *Pharmacokinetics and Cancer Chemotherapy*, Workman P, Graham MA, Eds. New York, Cold Spring Harbor Laboratory Press, 425-442.

Tong GL and Lien EJ (1976). Biotransformation of drugs: quantitative structure-activity relationships for barbiturates, tertiary amines, and substituted imidazoles. *J. Pharm. Sci.* 65: 1651- 1654.

Tterlikkis L, Ortega E, Solomon R, and Day JC (1977). Pharmacokinetics of 6-mercaptopurine. *J. Pharm. Sci.* 66: 1454-1457.

Von Hoff DD, Clark GM, Stogdill BJ, Sarosdy MF, O'Brian MT, Casper JT, Mattox DE, Page CP, Cruz AB, and Sandbach JF (1983). Prospective clinical trial of a human tumor cloning system. *Cancer Res.* 43: 1926-1931.

Vorontsov IN, Greshilov MM, Belousova, AK, and Gerasimova GK (1980). Mathematical description and investigation of the principles of functioning of the folic acid cycle. *Biokhimiya* 45: 83- 97.

Wagner JG (1993). Pharmacokinetics for the Pharmaceutical Scientist, pp.1-316. Lancaster, PA, Technomic Publishing Co.

Wagner JG, Aghajanian GK, and Bing OHL (1968). Correlation of performance test scores with tissue concentration of lysergic acid diethylamide in human subjects. *Clin. Pharmacol. Ther.* 9: 635-638.

Wang X and Ohno T (1995). Typing of MEIC chemicals according to their toxicokinetic modes of action by lactate dehydrogenase release assay. *In Vitro Toxicology* 8: 55-64.

Watari N, Sugiyama Y, Kaneniwa N, and Hiura M (1988). Prediction of hepatic first-pass metabolism and plasma levels following i.v. and oral administration of barbiturates in the rabbit based on quantitative structure-pharmacokinetic relationships. *J. Pharmacokinet. Biopharm.* 16: 279-301.

Webb JL (1963). General principles of inhibition. In: *Enzyme and Metabolic Inhibitors*, Vol. 1, New York, Academic Press, 1-949.

Weiniger D (1988). SMILES: a chemical language and information system. *J. Chem. Info. Comp. Sci.* 28: 31-36.

Welling PG (1986). Physiological pharmacokinetic models. In: *Pharmacokinetics: Processes and Mathematics.* Washington, D.C., American Chemical Society, 241-254.

Werkheiser WC, Grindey GB, and Moran RG (1973). Mathematical simulation of the interaction of drugs that inhibit deoxyribonucleic acid biosynthesis. *Mol. Pharmacol.* 9:320-329.

White JC (1986). Use of the circuit simulation program SPICE2 for analysis of the metabolism of anticancer drugs. *Bull. Math. Biol.* 48: 353-380.

Wipke WT, Ouchi GI, and Chou JT (1983). Computer-assisted prediction of metabolism. In: *Structure-Activity Correlation as a Predictive Tool in Toxicology: Fundamentals, Methods, and Applications.* Goldberg L, Ed. New York, Hemisphere Publishing Co., 151-169.

Workman P (1993). Pharmacokinetics and cancer: successes, failures and future prospects. In: *Pharmacokinetics and Cancer Chemotherapy.* Workman P and Graham MA, Eds. New York, Cold Spring Harbor Laboratory Press, 1-26.

Wright BA (1973). *Critical Variables in Differentiation.* 109 pp. Englewood Cliffs, NJ, Prentice-Hall, Inc.

Wrighton SA, Vandenbranden M, Stevens JC, Shipley LA, and Ring BJ (1993). *In vitro* methods for assessing human hepatic drug metabolism: their use in drug development. *Drug Metab. Rev.* 25: 453-484.

Wyatt JC (1994). *Computer Methods and Programs in Biomedicine* 43: 283-291.

Yacobi A, Skelly JP, Shah VP, and Benet LZ, Eds. (1993). *Integration of Pharmacokinetics, Pharmacodynamics, and Toxicokinetics in Rational Drug Development*, 270 pp. New York, Plenum Press.

Yamada Y, Ito K, Nakamura K, Sawada Y, and Iga T (1993). Prediction of therapeutic doses of beta-adrenergic receptor-blocking agents based on quantitative structure- pharmacokinetic/pharmacodynamic relationship. *Biological and Pharm. Bull.* 16: 1251-1259.

Yasuda SU, Schwartz SL, Wellstein A, and Woosley RL (1993). The integration of pharmacodynamics and pharmacokinetics in rational drug development. In: *Integration of Pharmacokinetics, Pharmacodynamics, and Toxicokinetics in Rational Drug Development*, Yacobi A, Skelly JP, Shah VP, Benet LZ, Eds. New York, Plenum Press, 225-238.

Yates FE and Kugler PN (1986). Similarity principles and intrinsic geometries: contrasting approaches to interspecies scaling. J. Pharm. Sci. 75: 1019-1027.

Zaharko DS, Dedrick RL, Bischoff KB, Longstreth JA, and Oliverio VT (1971). Methotrexate tissue distribution: prediction by a mathematical model. *J. Natl. Cancer Inst.* 46: 775-782.

Zaharko DS, Dedrick RL, Peale AL et al. (1974). Relative toxicity of methotrexate in several tissues of mice bearing Lewis lung carcinoma. *J. Pharmacol. Exp. Ther.* 189: 585-592.

Ziegler DM (1993). Recent studies on the structure and function of multisubstrate flavin-containing monooxygenases. *Ann. Rev. Pharmacol. Toxicol.* 33: 179-199.

Appendix

COMPUTATIONAL PHARMACOLOGY SOFTWARE SUPPLIERS

The attached list gives names, addresses, telephone numbers, and (where available) 1995 prices of the software discussed in the text. In most cases these are commercial products; a few are privately distributed programs that have stood the test of time and of multiple users. In each case a literature reference is given that documents the program or provides an example of its application. Unless otherwise indicated, these products are for MS-DOS computers.

ADAPT II
Ref.: D'Arginio and Schumitsky, 1979
Biomedical Simulation Resource
University of Southern California
Los Angeles, CA 90089-1451
Phone: (213) 740-0839
FAX: (213) 740-0343

BioCHIMICA
Ref.: Jackson, 1995
Agouron Pharmaceuticals, Inc.
3565 General Atomics Court
San Diego, CA 92101
Phone: (619) 622-3072
FAX: (619) 622-3299

CELLSIM
Ref.: Donaghey (1986)
Industrial Engineering Dept.
University of Houston
Houston, TX 77004

CLINICAL TRIALS
DESIGN PROGRAM
Biosoft
P.O. Box 10938
Ferguson, Missouri 63135
Phone: (314) 524-8029
Price: $299.00

CLOGP	Daylight Chemical Information Systems 18500 Von Karman Avenue, Suite 450 Irvine, CA 92715 Phone: (714) 476-0451 FAX: (714) 476-0654
COMBO	Bunow and Weinstein, 1990 Address listed under MLAB
COMPACT	Computer Optimized Molecular Parametric Analysis of Chemical Toxicity Ref.: Lewis et al., 1990 David Lewis School of Biological Sciences University of Surrey Guildford, Surrey GU2 5XH, UK
DEREK	Toxicity prediction Ref.: Sanderson and Earnshaw, 1991 LHASA UK, Ltd. School of Chemistry University of Leeds Leeds, LS2 9JT, U.K.
DOSE EFFECT ANALYSIS SOFTWARE	Ref.: Chou and Talalay,1984 Biosoft Address listed under Clinical Trials Design Program
EXCHEMO	Ref.: Jackson,1992 Included with the above reference
HAZARDEXPERT	CompuDrug North America, Inc. P.O. Box 23196 Rochester, NY 14692-3196 Phone: (716) 292-6830 FAX: (716) 292-6834
META	Ref.: Klopman, 1995 See information on MULTICASE
METABOLEXPERT	For MS-DOS and VAX/VMS Ref.: Darvas, 1987 CompuDrug North America, Inc. Address listed under HAZARDEXPERT Price: about $10,000.00

MKMODEL   Ref.: Holford, 1990
                   Biosoft
                   Address listed under Clinical Trials Design Program
                   Price: $499.00

MLAB   Civilized Software, Inc.
                   7735 Old Georgetown Rd., Suite 410
                   Bethesda, MD 20814-6130
                   Phone: (301) 652-4714
                   Price: $1495.00

MULTICASE   Ref.: Klopman, 1992
                   Biofor, Inc.
                   P.O. Box 629
                   Waverley, PA 18471

MuPHARM   Ref.: Saunders et al., 1989
                   IRL Press
                   Eynsham, Oxford, England

Mw\Pharm   Europa Scientific Software Corp.
                   14 Clinton Drive
                   Hollis, NH 03049-6595
                   Phone: (603) 595-7415
                   FAX: (603) 889-2168

NONMEM   Ref.: Beal and Sheiner, 1989; Sheiner and Grasela, 1991
                   School of Pharmacy
                   University of California
                   San Francisco, CA 94143

ORACLE   Oracle Corporation
                   500 Oracle Parkway
                   Redwood Shores, CA 94065
                   Phone: (415) 506-7000
                   FAX: (415) 506-7150

PCNONLIN 4.2   Ref.: Metzler and Weiner, 1992
                   SCI Software
                   2365 Harrodsburg Road, Suite A-290,
                   Lexington, KY 40504-3399
                   Phone: (606) 224-2438

PHARM/PCS   Ref.: Tallarida and Murray, 1987
                   MicroComputer Specialists
                   P.O. Box 40346

Philadelphia, PA 19106
Phone: (215) 625-9022

Ph/EdSim

Europa Scientific Software Corp.
14 Clinton Drive
Hollis, NH 03049-6595
Phone: (603) 595-7415
FAX: (603) 889-2168

PHYSIOPLEX

Address inquiries to the author
Address listed under BIOCHIMICA

P-PHARM

SIMED S.A.
9-11 Rue G. Enesco
94008 Créteil Cedex, France
Phone: (+33-1) 45 13 13 25
FAX: (+33-1) 43 99 05 88

pKalc

BioSoftware Marketing
4151 Middlefield Road, Suite 109
Palo Alto, CA 94303-4743
Phone: (800) 456-4276PROGNOSYSHealth Designs,
    Inc.
183 East Main Street
Rochester, NY 14604
Phone: (716) 546-1464
FAX: (716) 546-3411
Generates prediction modules for TOPKAT

PROLOGD

CompuDrug North America, Inc.
Address listed under HAZARDEXPERT

SAS/PC

SAS Institute, Inc.
Cary, NC
SAS/STAT Guide for Personal Computers
Version 6, 1987

SIMUSOLVE

Ref.: Rey and Havranek, 1992
Mitchell and Gautier Associates, Inc.
200 Baker Avenue
Concord, MA 01742
Phone: (508) 369-5115

SIPHAR/win

Address listed under P-PHARM

STATISTICA   StatSoft
                                    2325 E. 13th Street
                                    Tulsa, OK 74104
                                    Phone: (918) 583-4149
                                    FAX: (918) 583-4376
                                    DOS version: $795 (Quick STATISTICA: $295)
                                    Windows and Mac versions also available

STELLA II   General purpose dynamic modeling package
                                    High Performance Systems, Inc.
                                    45 Lyme Road, Suite 300
                                    Hanover, NH 3755-9902
                                    Phone: (603) 643-9636

SYNFIT   Ref.: Greco et al., 1990
                                    Not yet commercially available
                                    Address inquiries to:
                                    Dr. William R. Greco
                                    Roswell Park Cancer Institute
                                    Buffalo, NY 14263
                                    Phone: (716) 845-8641

TABLECURVE   Jandel Scientific
                                    San Rafael, CA

TOPKAT   Ref.: Talley, 1993
                                    For MS-DOS and VAX/VMS
                                    Health Designs, Inc.
                                    183 East Main Street
                                    Rochester, NY 14604
                                    Phone: (716) 546-1464
                                    Price: about $10,000.00
                                    Note: Now know as "METABOLITE"

VIROCHEM

Ref.: Jackson, 1996
Address listed under BIOCHIMICA

Xenobiotic Metabolism
database

Ref.: Gifford et al., 1994
Molecular Design, Ltd.
2132 Farallon Drive
San Leandro, CA 94577
Phone: (800) 635-0064

Index

C